The Four Philosophies of Lean

The Four Philosophies of Lean

Maintaining a Customer-Focused Culture Every Day at Work

Robert Corbitt and Cory Bronger

Routledge
Taylor & Francis Group
A PRODUCTIVITY PRESS BOOK

First published 2022
by Routledge
605 Third Avenue, New York, NY 10158

and by Routledge
2 Park Square, Milton Park, Abingdon, Oxon, OX14 4RN

Routledge is an imprint of the Taylor & Francis Group, an informa business

ISBN: 978-1-032-04821-5 (hbk)
ISBN: 978-1-032-04819-2 (pbk)
ISBN: 978-1-003-19478-1 (ebk)

DOI: 10.4324/9781003194781

Typeset in ITC Garamond
by KnowledgeWorks Global Ltd.

Contents

Preface

Anyone in the manufacturing industry in today's time is familiar with the philosophies and practices of Lean. What began with Toyota and the automotive industry, has rapidly spread throughout most of the manufacturing world. One does not have to look hard to find a book or manual on Lean, the Toyota Production System, and its history. Every company is striving to mimic the success that Toyota has established. Many leaders of these company's turn to books to start down their Lean journey and while these books give the method, the understanding is all too often the failure point in the sustained execution of the implementation of the "method". This understanding only comes thru coaching during the practical application or as most in the industry say on-job development (OJD). So, why are companies unable to achieve the desired results?

I was recently asked a similar question and my reply was simple because Lean is NOT a "thing" you go do, but a way of life. It is simply how the business is run every day! Coming from a 50-year combined journey in automotive, agriculture, industrial, and aerospace manufacturing, we will explore the roadblocks that hinder Lean deployment and sustainability. In the upcoming chapters using our life experiences, we will do our best to explain why companies, although desiring Lean results, fail at achieving much success in implementing the necessary systems and concepts that Lean requires.

This is chapter one: First, we need to understand what Lean is. My first introduction to Lean was as a team member on the assembly line at Toyota Motor Manufacturing Kentucky. It was a new concept for me and I remember asking what is Lean? The answer I got back was "what we do every day". That answer left a lot to be desired and looking back now is one of the very reasons Lean fails today. At that time what we did every day at Toyota just felt normal and seemed like common sense. What I found out later is that, what made sense to me was not the same for others. They may have had different challenges and may have not had the experiences that I had. Understanding why we do something is just as important as the how we do it in relation to a process/task/job. So, what is Lean? Lean is a systematic approach to identify, visualize, and eliminate waste by utilizing employee development and continuous improvement in all phases of manufacturing products and provided services. The acronym for LEAN does not mean Less Employees Are Needed. That is a misconception among many due to its targeted efficiency gains. Whether we realize it or not, we all use some form of Lean in our everyday lives to improve our tasks, free up our time, minimize waste, and save money. Pretty simple, right? It should be but the path to implementing a Lean production system has many obstacles and that is what we are here to discuss, reflect upon, and hopefully give clarity to what all too often does happen as opposed to what should be happening.

Let's digest that "What we do every day" response. Something we need to consider with that phrase in mind is, that what team members do every day at Toyota is not what everyone else in the industry does. Toyota team members are immersed in a Lean culture or as one might say "immersed in a way of life, running the business" when they walk in the door. It is for that very reason when we left Toyota and entered other companies out in the industry, we were to say the least shell shocked. Every day running the business practices were not so common among these other companies. Our workplace was clean, orderly, and standardized to the degree that every desk was standard to support the elimination of time spent hunting down needed items. No matter where you were in the plant, you knew where to get whatever you needed. Everything was clearly labeled and communicated for all employees to be easily integrated into any area new to them. By having this type of infrastructure, it simplifies the onboarding of new employees. Problems were not dismissed or avoided but encouraged to be brought to attention and addressed by a systematic approach of *kaizen*. Teamwork was emphasized and accountability started at the top then moved down the ranks. The Toyota culture encourages workers to actively seek continuous improvement in their processes, surroundings, while also empowering them to implement needed changes. Very few new employees have the understanding that this Lean thinking applies to almost everything that happens in their daily work. Simply put, this understanding that the culture we were exposed to, day in, day out at Toyota for over a combined 33 years created a new normal from other work environments outside the Toyota walls. The greatest challenge we faced stepping into a new role with these different companies wasn't what our new role required but rather seeing Lean so poorly implemented or not utilized at all and trying to break through the barriers to get results. Often the hardest obstacle to overcome was simply arrogance; among workers, leadership, and sometimes even ourselves. This arrogance stems from experience and comfort within the system in which they have operated. That experience can be both an asset but more importantly a liability. The asset portion is knowing your job and the history of the role and company. While the liability means you have limited exposure and creates a sense of ridged thinking. All too often it is very hard to "See" what the next level of good would look like. Toyota often prefers new people vs an experienced person, their motto is "We Build People Before We Build Cars"

When you know "everything about everything", you stop learning!
When you Stop Learning you Stop Growing!

Our journey in Lean started with Toyota but it wasn't until we left, we began to want to learn more about the Lean process. Understanding that not everyone is on the same level of thinking, even leadership is challenging. The lack of understanding connects to the obstacles in the implementation of Lean. My meaning is Knowledge does Not equal Understanding! So, clearing the air between Knowledge "The Method" vs Understanding "The Intent or Intended Purpose" presents quite the challenge for many leaders. Think about a STOP sign and what our Knowledge of a stop sign is, "The Method" is that when you come upon one, you stop your car/bike/walking. But what is the true understanding "The Intent" is to save lives. This deeper understanding of Lean principles does not always transfer out to other companies. While Toyota invests in their people and empowers them to act, this is not a common practice throughout the industry and is something we took for granted. The manufacturing industry, wanting Toyota's success, started poaching Toyota workers, creating a niche for Lean consultants. This has become a mixed bag successes and failures. Some folks have implemented Lean with tremendous success. While on the flip side some companies and people have been left with a bad taste in their mouths, so much so, that although our expertise is sought due to our Toyota background, we have often been coached on not emphasizing it. Time and again leadership has hammered away at their workforce on how great Toyota is then brought in new individuals to roll out Lean systems only to revert to old habits due to never having come to a full understanding of the implementation. This has a very adverse effect on the workforce moving forward. It sets the stage to become the "flavor of the month". When you've been surrounded by best practices, developed to see problems and solve them, it's tempting when strolling into a new company to start pointing out the issues with an inflated attitude. This is one of the quickest ways to shut down engagement. We want our audience to be approachable and coachable, but we must realize that we, ourselves, must also reflect the desired behavior of being approachable and coachable. It is my opinion that when people keep these two words in front of their thoughts, it sets the stage for stronger two-way communication. So, what does it really mean to be approachable and coachable? I list these words in this order because I believe my demeanor "approachable" must be reflected in my body language and behavior to set a comfortable, inviting atmosphere. While being "coachable" means I will actively listen and internalize the conversation before making a decision. As we practice using these terms, our leadership skills will grow, as will the number of co-workers seeking alignment or advice before moving forward.

This goes back to the "How we do something, is as important as the why we do the thing. So, how we engage our people directly affects the outcome. People overall are resistant to change, to begin with. When we enter a new environment and start attacking the problems and making changes without engaging employees respectfully, gaining their buy-in, and developing their skill and capability to recognize problems and make changes going forward, we eliminate a key factor of Lean.... SUSTAINABILITY. There are several companies that have hired Lean consultants that had their ship righted only to show a steady decline after the consultants left. This tells us that Lean was not implemented correctly. That's a hard pill to swallow but our arrogance is the first hurdle we must get past in order for Lean to be successful. The best example of this is a conversation that was had with a former plant manager. During the conversation, the plant manager bragged about how when he left one site of the company to go to another site, the site he left fell apart. In his mind, he was paramount to the company's success and on the surface, one could say that. However, had he truly followed Lean and implemented value-added processes, he would have developed his people and the company wouldn't have skipped a beat. "If the student hasn't learned, then the teacher hasn't taught". The first reason why Lean fails is due to arrogance. We must learn to be humble. If we can accept that and hold ourselves accountable, then we can begin to tackle the other problems facing Lean.

Acknowledgments

Robert and Cory have been on this journey together for the better part of their Lean journey and have had many of the same leaders that have impacted their lives in how they lead and teach. We would like to say thank you to several instrumental people. Dinesh Vasandani for being there at the beginning with the mind of a teacher to kick start our Lean journey. Dave Gibson for giving us both opportunities to learn and apply what we have learned. Dave had and was a library of information that he translated into real hands-on learnings. Dave handed me my first book and said "start here and come get your next one soon". Patrick Ford for being an incredible leader with a strategic mind for launching Lean thinking.

From Cory to his wife:

To Renee Bronger, my wife that has been nothing less than my biggest advocate, always supporting me at the highest level. Thank you for picking me up on those tough days and helping me to see all the good in each of those challenging days. For allowing me to chase the pursuit of growth for myself and those around me.

From Robert to his wife:

To Trish Corbitt, my better half, my sounding board, and my biggest supporter in helping me achieve so much in my life. I want to thank you for being there for me when my work life was not allowing me to always be there with you. You are the best thing to ever happen to me and I am so happy we are in this life together building our family and love.

About Authors

Robert Corbitt has 27 years' experience in Continuous Improvement in manufacturing. Sixteen of those years were at Toyota Motors Manufacturing in Georgetown, KY. Certified A3 trainer and Lean coach. He attained a bachelor's degree from Northwood University in Manufacturing Management later in his career, never too late to learn. Several President's awards for Operational Excellence from Ingersoll Rand. He has had roles from a team member on the line assembling cars to a role as a Global CI manager supporting many plants across the globe in their Lean journey. He was able to launch AGCOs Production System in its new business unit Grain Systems Incorporated. He has also worked with numerous departments in the efforts of continuous improvement, such as purchasing, sales, operations, materials, accounting, engineering, and so on. He has worked with a diverse customer base related to (Automotive, Agriculture, Industrial, and Tear 1 Automotive). He has also worked in many different countries with different cultures and beliefs. He has also owned and operated his own Lean consulting business. He is a teacher and student at heart for the philosophies of Lean and people development. Always sharing and learning day by day.

Cory Bronger is a Senior Lean Practitioner at The Boeing Company and the owner and operator of CDB Consulting. His life story is one of constant improvement. Quitting high school at 16, he obtained a GED and welding certification. His first job as a welder/fabricator for United Industries, a railcar company, often required him to design and fabricate tools in order to complete his work. This is where his enthusiasm for problem-solving first began. Changing direction in 1998, he began a 17-year long development journey with Toyota Motor Manufacturing in Georgetown, KY. Starting as a team member in assembly, he worked in many roles and on several different project teams, assisted in the startup of the Toyota Mississippi factory, and holds a US patent for automated dolly assemblies. After leaving Toyota, he worked as an Operational Excellence Change Agent with Ingersoll Rand where he was awarded two President's awards and a chairman's award for his role as the lead site trainer in the deployment of a Leadership Development Program across the United States and China. He started CDB Consulting in 2017, consulting for companies in the United States, Mexico, and Belgium on their Lean systems. He is the epitome of Lean, overcoming problems and challenges in his own life while constantly improving. Combining his professional and personal life to mentor others is his passion.

Chapter 1

Lean Thinking Philosophy

Why does Lean fail? Why can't we understand how to do Lean? In the coming chapters, we will use our life experiences and share examples of success as well as where people have stumbled. This story is coming from a 50-year combined journey in manufacturing in the worlds of automotive, agriculture, industrial, aerospace, and so on. I would always tell people that Lean was ingrained from day one at Toyota. Not all of us embraced it 100%. I can tell you the ones that did always wanted more and drove Toyota to the level it is today.

The other problem we faced later after we left Toyota was what we thought was common sense, which was not so common among others. This was no disrespect to others. It was what we had been exposed to day after day at Toyota for over 16 years; we had been developed to a deeper way of thinking. Not everyone has this Lean thinking applied to almost everything that happens in their daily work. It was important for us to understand that not everyone is on the same level of thinking, even leaders. That is where our journey started in wanting to learn more about the Lean process. The other thing we have learned over the years is that Lean is not just about cost, nor is it an acronym meaning **L**ess **E**mployees **A**re **N**eeded. Lean is bigger than merely a cost-saving mentality; it is a philosophy, a way of thinking in everything we do, a system of principles to be our best and strive for more in everything we do. When we get our people on this level of thinking, we can accomplish anything. The growth of our people is far more valuable than we can even imagine.

DOI: 10.4324/9781003194781-1

Customer First

Lean has four areas of focus when it comes to philosophy: Customer First, Respect for People, Go and See, and Continuous Improvement. Customer First is an area that needs to be developed by many people. When we say customer first, we need to be thinking of our customers in every portion of the production process. How our actions will affect our customers. We need to protect our customers from all internal and external challenges. In the Lean philosophy, the customer is not defined as the "purchaser" or "end-user" of the product, but rather everyone in the production system. Developing our people to understand their role and its importance to deliver a quality portion of the product to the next step in the process is key for success. When we design a product, a company needs to think in terms of "Brand Commitment". Are we providing what the customer wants? Are we delighting our customers by providing value? These are questions to be understood. What customer wants and can we make it better so we can delight our customer and gain brand commitment? Is it the cost that the customer is willing to pay? Can we drive the cost down to gain more market share? When we manufacture our product, are our processes stable to protect our customers from shortages? We should have a focus on "Jidoka" (built-in quality) to our processes so we need less inspection at the end. This protects our customers from defects and keeps our cost down for our customers, the shareholders. Yes, we need to also recognize our shareholders as a customer. How often does a company launch a new product only to run the cost over budget as well as struggle to meet our customer demand and deliver in the desired timeframe? As a result, we often support by sending people to our customers to correct or sort the defects. All of this leads to a higher cost with a lower competitive advantage and lower customer confidence.

How many times have we started a project and have not truly understood our customer's needs, or took time to list the needs and ensure everyone understands them? I remember being in a workshop and asking the question about a third of the way through, what are we really trying to accomplish? What are the real problem and the expected outcome? I believe all projects and leadership training should start with customer-first thinking, and how we can make the customer happy. A clear vision of our goal and the problem to be solved is the best way to get everyone on the same page to succeed. A3 thinking creates by default a mechanism to support the road map of alignment. If you start a project with a strategy A3, you will clearly explain the problem by showing the current and future states. It explains to others the thought pattern that drove us to a conclusion, it manages the project visually, and it tracks our success and what we learned from this journey. The organization of the A3 information would give structure for all to learn and understand. We will further discuss this concept in the A3 Thinking section.

Even on the micro level, we need to think about our customers. The next process is a customer. Companies often miss an opportunity to gain a competitive edge by educating their workforce about a "Don't make it, Don't take it, Don't pass it on" mentality, and each downstream process is "your" customer. This can start leading to a sense of ownership by operators. If I am the next customer, some questions may start to arise. How do we present the product to the next process/customer? Are parts presented in a way that is easy to pick up? Are parts clean of debris from the current process? Will "I" (the operator) give my next process/customer a product that will reduce waste in their process? Do we take pride in our work so defects don't pass to the next process? We should not expect to make defects, and so we should not expect to receive defects. The complete motto I say a lot is "Don't make it, don't take it, don't pass it on". Every process should be a stop gate for defects. Stop the process and correct the problem, and protect yourself and the customer.

Respect for People

The second Lean philosophy is respect for our people because they are our most valuable resource. Think about it from this perspective, in our business only people can take ownership of their development and drive it to the next level. A machine or product will only be as good as its intended purpose and the people that contribute to it. The same ideology can be said for the development of our leaders/people. When it comes to leaders, we want to develop them to be their best by giving them stretch assignments to be better. Teaching them respect for others in each aspect of their role and responsibility. These "Seven Guiding Principles of Toyota" were a part of the business model during the early days of the company, say around the 1950s. As you read down them, you will find a host of values and ideologies that lead to respect for people. The list is as follows:

1. Honor the language and spirit of the law of every nation and undertake open and fair business activities to be a good corporate citizen of the world.
2. Respect the culture and customs of every nation and contribute to economic and social development through corporate activities in their respective communities.
3. Dedicate our business to providing clean and safe products and to enhancing the quality of life everywhere through all of our activities.
4. Create and develop advanced technologies and provide outstanding products and services that fulfill the needs of customers worldwide.
5. Foster a corporate culture that enhances both individual creativity and the value of teamwork, while honoring mutual trust and respect between labor and management.
6. Pursue growth through harmony with the global community via innovative management.
7. Work with business partners in research and manufacture to achieve stable, long-term growth and mutual benefits, while keeping ourselves open to new partnerships.

Source: https://blog.toyota.co.uk/the-7-guiding-principles-of-toyota

After reading the principles, one cannot ignore the differences of cultures and how we value or hold ourselves accountable to those words. On my first trip to Japan, I found it amazing how clean the country was considering the population size, but during my stay, I saw why and how. In Japanese culture words like honor and integrity are made apparent through the behavior of the people. I remember watching this elderly lady out on the street in front of what appeared to be a random convenient store plucking all the dead flower buds off the plants and sweeping the area. For me, this behavior reflects many of the keywords that are written into the guiding principles from so many years ago.

A way for leadership to show respect is to have a strategy to develop people at all levels to become stronger leaders, that is, building skill and capability in how to identify, visualize, and eliminate problems, and to be able to coach those "ah-ha" moments. Coaching is a developed skill that should be taught through question learning or "Socratic leadership". This method of development can be very powerful. But one must be cautious not to ask questions in a way that can be viewed as condescending. We want to ask questions that get our people to begin to engage with their thoughts on how to come up with a solution. We may need to guide the conversation with our questions and support learning. A good coach will seek to understand what is needed. They will be open to different ideas and always follow up to help guide the learning process.

Attack the process, not the person. Most of the time if a person is disgruntled, unhappy, or complaining, if we dig deeper, we will find that there will be a problem with their process driving the undesired behavior. This would be an example that we may not be promoting the respect for people philosophy. Help understand our goals and how we can reach the company targets. How we protect our customers. Understand that they have customers on many levels. How to be servant leaders is part of this thinking. This type of thinking does not stop with leadership; it rolls down to the people that add real value to the processes we perform and the products we produce.

Another way for respecting our people is when we improve a process. We should always try to put ourselves in the team members' shoes. It would be about making their life better while they work based on the mindset of Safety, Quality, Delivery, and Cost. Even the micro details matter such as lighting, ergo mats, etc. Can the operator build quality into the standardized work, so no defects flow out to the downstream process? Has the process been simplified so we can train a new person fast and with minimal technical skill development? Is the process loaded with as much value-added work as possible? We want to avoid non-value-added work in our processes, this is wasting our people's time and for many people, it creates boredom and thus is not respecting our people. Is the ergonomics good and can I make it better to reduce the fatigue, the cost of injuries? Are our processes balanced? Do we have the right amount, of resources to support the team members from a ratio viewpoint? In terms of managing an area or a team, the ratio of team members to team leaders should be relative to the level of the position. My point is a lower leadership position such as a team leader should not be expected to manage 10 or 20 team members. This should be closer to a 1 to 5 ratio, and as a person moves up the ranks that is when their coverage of management would grow. Over-loading a lower-level leadership position would be a deterrent to moving up or possibly detrimental to a person's mental stability and so disrespectful to our people. Climate can be one item that is easily overlooked. Do we have the proper air movement for when summer gets here or heat for when winter gets here? That is one that I saw missed the most. The key to respect for people is doing the right thing for your people and they will respect and follow you as a leader in the future. Even during times of reducing operations when we make sure the above was done right, we had happier employees and greater success in the change, which protected our end customers and shareholders.

Go and See

Go and see on the shop floor, Go to the Gemba (where the work is happening) as one might say, are about the third Lean philosophy. You cannot solve your problems from your desk or a meeting room alone, so get your boots on. The shop floor is changing all the time and we need to witness what is happening while it is happening. We have always used the stand in a circle or better known in the manufacturing industry as the Ohno circle method to help see the "real". This methodology is about observing the process, looking for waste, and any item that impacts the process. When you go to the Gemba, you should be there to observe first, while considering what questions need to be asked to clarify the current situation. It is necessary to make sure the employee understands what you are doing and that they will be involved once you observe the process or task. As we watch, remember we should be trying to understand the process as of today and what are all the components that can impact it. Material and information flow, human movement, part orientation and fit condition, equipment performance, overall environment, even down to the facial expressions of the operators. If the process has standardized work, it would be a good time to review and compare what you see is really happening. This way you can start to see if there is a gap between the standard and the current condition. Go into the environment with an open mind, drop, or suppress your preconceived thoughts. When you do this, you can start to see the "real". Remember I mentioned three keywords earlier, identify, visualize, and eliminate. With those words in mind, you can then start to develop questions to go deeper into the problem. You will be better equipped to filter out the noise from complaints to what is real. There have been so many times people complain about people, they target the person before the process and the real problem is that the process is broken, and it is driving a bad behavior. This leads back to being able to identify the root cause of the problem versus chasing a symptom of the problem.

I have found so many times that going to the floor or where the work is happening with my team, we get a better understanding. Questions that came up in the conference room are answered or additional questions are raised. I have also found I get to know our people so much better and learn so much more about the other areas that we can support them to do their daily work in the allotted timeframe. This has been a great way to develop relationships and change the lives of people that work so hard for us all to be successful. I have found over the years that respect can flow both ways, but I must start with myself to improve the connection from bottom to top. People may respect the position while not respecting the person. One way this can be changed is by being engaged and being an incredible listener. Look for improvement opportunities to help make their process better, not just correct the original problem. Think about it, let's say we are trying to correct a defect coming from an upstream process, and while we are watching we learn that the ergonomics can be improved with a simple move. Correct both and we exceeded what was expected from our customer the team member. It's a win, win. The defect is corrected and the team member goes home at the end of the day not as sore from the process. We don't have to spend thousands of dollars to make people happy, just show respect in the little things. Listen, be clear in what you are wanting to do, get buy-in, ask for their opinion, challenge people to exceed, deliver on what you say, when you fall short be humble, and make it right. We should also address the corrections to the process with a sense of urgencies. The faster we make the changes the faster we get our processes to normal, and our people feel valued. It also shows the importance of having a better process for the customer and our people that do the work. Just think about it; why would you want a problem to go on for one more second when we know the solution?

Continuous Improvement

Continuous improvement is the fourth part of Lean philosophy. We must always strive to improve. In the business world the ground, under your feet is never standing still for any length of time. If we are not working to make improvements our competitor will capture market share. Continuous improvement of price and a product with better quality leads to sustainability, which supports a better place to work. Companies should always be looking at ways to raise the bar and bring their people up with it.

This thinking will lead to driving our cost down by empowering our people to be better problem solvers. Continuous improvement is often simple incremental innovations in our business that culminate into big changes. One of the keys points with continuous improvement is to develop standardized work for each process. This will cement the changes, so they are not lost through “tribal/travel” knowledge. We will explain more later. Standardized work is also a foundation piece to build upon to the next level. Continuous improvement is not just for the operations on the shop floor, it is for all processes in your business. By driving continuous improvement, we will give our customers a better product with repeatable, sustainable, quality processes.

Continuous improvement should be happening so often that maintenance would want to put casters on every piece of equipment so that it is easier to move around or adjust on the fly. Even after an improvement has been made, we could always go back to a process and ask are there more improvements to be made? What waste did we leave with the current technology? Did we ensure the process is building in quality every time in cycle time? Will further studies benefit both the operator and the company? And in reflection, what did we learn to streamline the next improvement? This is not always thought about. I remember when I started at Toyota, they would take two weeks of lost production to change from one major model change to the next new model. When I left Toyota the model change process took about 30 minutes on the line, and the most we lost is a few minutes of work time. We had continued to improve the process over many years to get to this level of ability. This is the key to success in continuous improvement. Reflecting on what we learned and plan, plan, plan for the next evolution of standardized work. Driving continuous improvement and sustaining it is important to succeed in business. How do we do this? We need to track our success to make sure that it is sustained. When improvements are being made data should support the decision to move in a given direction, but often leaders with lengthy experience make decisions based on what they know or think they know.

All four Lean philosophies are tied together if we are thinking in the right way. With continuous improvement in our daily decision-making, we will drive customer first, respecting our people, and shop floor thinking. When we go to the shop floor or where the work is happening, we show respect by participating and interacting with our people, seeking to understand their challenges, and then removing the roadblocks. This will support driving cost down. Driving cost down will be a by-product of the behaviors and the way of thinking we are describing. Always attacking the process, not the person, is what we should be targeting when we are on the floor. Attacking the person is easier than identifying a gap in the process the person is performing. If we are in that type of mindset, we are showing respect to people.

Our goal in this book is to show you how these four concepts of this philosophy are tied to how we should be running our daily business. Each chapter should help connect dots along the way. Without this philosophy, we are not truly understanding the real meaning of Lean and is a major contributor to failure in Lean efforts. We are then just applying Lean tools and never growing past the point of seeing the method to understanding and realizing the full benefits.

Chapter 2

A3 Thinking

Why do we fail at solving problems? Why do we fail at implementing a strategy or project? Why do we have trouble understanding what we need to invest in equipment, project, or simply human capital? Many times, in our past, we thought we were solving a problem and all we did is attack the symptoms, allowing the issue to return again and again.

Just a fun fact, the term "A3" is actually just the size of the paper that people used years ago to tell their story. A3 is equal to an 11" × 17" tabloid size piece of paper. This size is used and is needed to be able to capture all necessary information to solve a problem. A typical A3 would have eight to nine boxes/steps. Step one: Define the Problem, step two: Breakdown the Problem, step three: Target Setting, step four: Root Cause Analysis, step five: Solution Approach, step six: Rapid Experiments, step seven: Implementation/To-Do List, step eight: Measure Results, and finally step nine: Reflections. These steps are key to help us develop deeper thinking to solve tough problems.

We have found in our past that not everyone understood the purpose of why we need to change and where we were wanting the company or team to go. When people don't understand the intent of why we are changing, there is a greater chance we will do it over again because we did not understand it the first time, or worse, revert back to what we were doing before with no learning. Our competitors are always trying to pass us up, and in business, the ground is always moving under your feet. If we don't have alignment and understanding to guide us on our journey, it would be very possible for our competitors to gain a competitive advantage over us, and that translates into a gain in our competitor's market share. So, the

DOI: 10.4324/9781003194781-2

equipment that doesn't meet our needs for the future is often purchased and sets the stage for the equipment or systems that create more non-value-added work. We often repeat these unfavorable decisions over and over because we didn't clearly understand the real problem and we can't articulate how we came to a given conclusion to justify the direction. We end up telling a story through power points that are often subjective in their conclusion at best, leaving our leaders lost in the review. We attempt to give clarity but not all understand our thinking and how we came to the conclusion. As we systematically think our way through and improve our A3 thinking skills, we often identify gaps in the information flow, this is what leads to the lack of clarity. When developing these A3 skills, four letters should come to mind that will support and guide us – PDCA (Plan Do Check Act/Adjust). Most companies have a diagram that correlates how the different phases of PDCA connect to a typical A3 document, this will show where you are in the document cycle.

There are three types of A3s that we have used in our past. Problem Solving, Strategy, and Proposal A3s. There are many books that go into great detail on this subject. We are just going to walk you through these documents and the process for basic understanding.

An A3 is a great way to tell a story on one sheet of paper. It follows the PDCA format to clearly show your thinking process. One way I have described the A3 is through the PDCA process. The Plan portion is like a charter and the breakdown or analysis of the thinking. The Do section is similar to project management. The Check section is there to ensure we have won, if not we then can act on what we learned. It is sometimes important for people to be able to relate to the purpose of each section and how they may have used those concepts. It is a document that will start driving people to speak the same language using common terminology. How many times have you been in a meeting and thought "how in the world did you come up with that as a solution"? How many times have we missed key information or processes that lead to not winning? You will find that an A3 will help us tell our story, capture the key items we need for success, and drive alignment with the team. Another note about an A3 is that many companies will develop a standard template A3 form to drive standardization in the company. I always caution people, "Don't let the form be rigid or literal". Let the information flow, not every item will fit neatly in a 2" × 3" box. It has been my experience that people often fall victim to this constraint because it is a company "Standard Template". What we are saying here is that the size of boxes should not constraint you from telling your story. In our past, we were trained on a blank 11" × 17" (A3) sheet of paper to develop us to be systematic thinkers with clear understanding. All A3s need to follow the PDCA cycle to tell your story.

Problem Solving A3

This can be an interesting subject for us. I cannot think of how many times I have been told "hey I problem-solve every day" only to identify what they are doing is firefighting. Then you have the ones that are really trying to solve the problem and the problem seems to always come back. Firefighting is always going to be needed because we must contain the problem at the point of detection. This is how we protect our customers from defect escapement. But firefighting often leads to people losing faith that anything can be fixed because the problems always come back. This person is not finding the root cause of the problem, but rather pouring some water on the fire until only ambers remain. And as we all know ambers lead right back to a fire. There can be many reasons for not finding the root cause, by using A3 thinking we will be better equipped to kill a problem at the source (refer to Figure 2.1). Finally, we have the person that says we don't have any problems. Everything is green on our boards and we are winning every day. This person's thinking may be the biggest problem to understand first. If we are winning all the time maybe we have low standards for success. Maybe we are not challenging ourselves to get better. We have seen people try to hide the opportunities to improve. There are many reasons for this, one being they don't want others to see that they are not as good as they seem. Others may have accepted that they can't correct the problem so they set standards so the problem is not a focus. Either way, we should always be thinking of continuous improvement in all that we do. That is how we should see ourselves winning every day. In this section, I will give you some highlights to support A3 thinking.

First things first, we need to understand how to scope a problem to solve. Solving world peace is not what we should focus on. If I am a supervisor and we have an On-Time Shipment metric that I am able to hit every day my target may be too low, we may want to lower the water as many people out in the industry say. Lowering the water means changing our target to drive higher performance. This will expose the rocks below the water surface. These rocks (problems) are typically independent problems that need separate focus to solve each one, in other words, we must reduce our

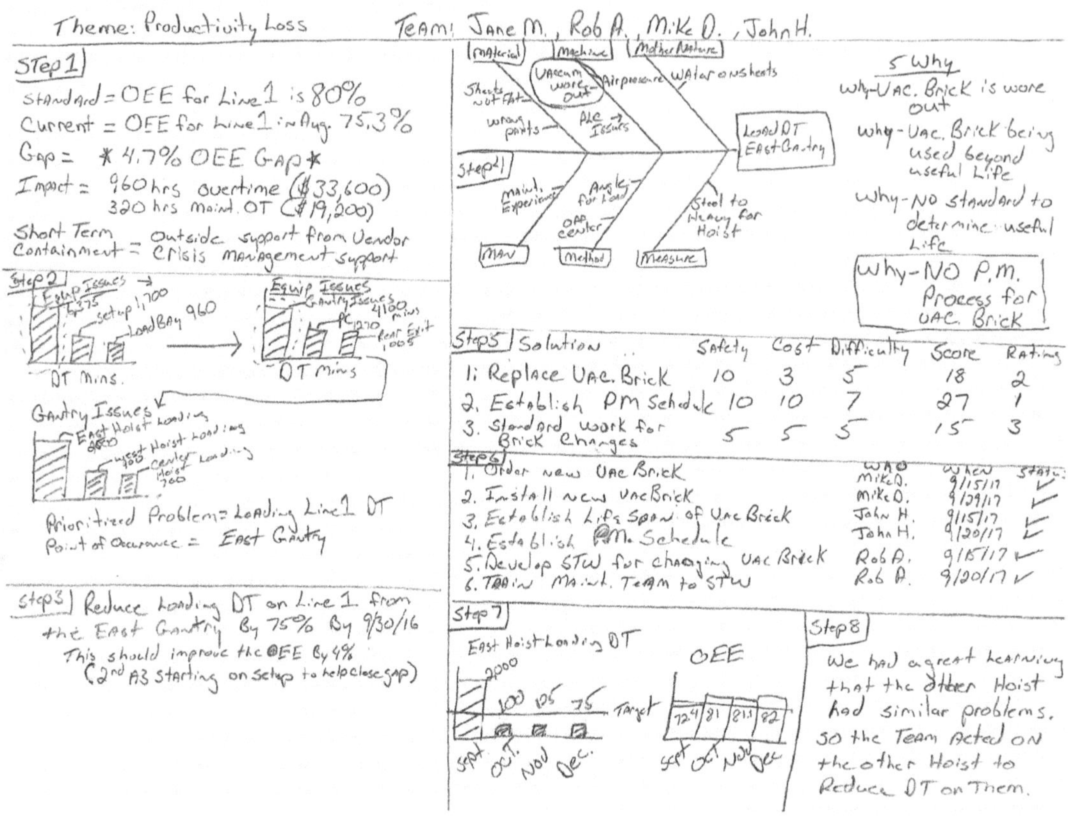

Figure 2.1 Problem Solving A3.

scope to focus on each. By reducing the scope, it will be easier to find the root cause of an issue and solve a problem forever. Just think about a metric that can be affected by several departments and in those departments, there can be many contributors to a problem. In the end, by narrowing the scope and applying lazar focus to eliminate each smaller problem, the bigger problem will take care of itself and all departments will reap the benefit of a higher-performing team.

How does an A3 tie back to a metric in your company? If you cannot tie it back to a metric, you may be spending your time on an issue that does not matter to the company and you will have a hard time getting support and buy-in. You should also have the current metric, so you know what is happening. Now that you know the metric you are trying to improve, you should have a standard or expectation of what you want to accomplish. As you define the difference between the current state and the target state of what you are going to work on correcting – that will be what we call the gap. We also want to explain the impact on our business from this gap. This helps others clearly see the reason why we need to work on this particular problem or simply work on this opportunity. Do not leave step zero or rather containment out of the mix by adding short-term countermeasures in place to protect our customers. Next, we will call step one, understanding the current state, the standard which leads to the gap and impact on our business. In our example, we see the gap to be solved is 4.7% Overall Equipment Effectiveness (OEE). This is creating a productivity loss and the site is trying to protect the customer overtime.

The next step (step two) is to take the data you should have from the gap and break it down to a smaller problem to be worked on. Driving a problem down small enough to fully understand it and kill the root cause of the problem. Breaking the larger problem down to very small problems makes it easier for the team to solve, we can't eat an elephant with one bite, we do so one bite at a time. Breaking data down into a multitude of Pareto charts helps narrow the focus. We also like to break the data down in different directions to see if it shows me something different. Using the data will help us see if something is sticking out, which leads us down a path. I have used this analogy in the past, data is like a hayfield and we are looking for the needle in it. The goal is to break it down until we have just a hand full of hay and the needle would be in that hand full. Take a large problem and break it down to something we can better understand and attack. Once we get to this point, we should go and see what is going on in the area that the data is telling us to go to. Go and see observation is a huge part of this process. We need to stop right here and make sure we are thinking like a detective, not like the supervisor from the area that feels they know the answer. We should ask questions about the smaller problem and look upstream and downstream. Spending an hour or more watching multiple cycles of a process to determine what is going on in the area is a good start. By breaking down your problem, then using the go and see approach, you should find the Point Of Occurrence (POO), which will lead to the Prioritized Problem (PP). You may even add other data or even a drawing to help explain how you got to this point in your A3. In our example, we broke the line downtime down to understand that the east hoist was having issues with loading the line.

The next one (step three) should be simple, draw the line in the sand and decide how much of the problem you would like to fix. This gives you a goal and also lets everyone know if you need to add to the problem solving by starting another A3 with another team. If the problem is broken down into a small problem, and you understand the POO and the PP, you should have a better understanding of how much you can win by. We always want to be clear at this point by stating: what we are going to do, to what, by how much, by when, and include how much we are closing the original gap by. In our example, we planned to reduce the downtime on the east hoist for loading by 75% within 30 days.

Next is step four, getting to the root of the issue, you need to think about all the things that could be part of the problem. One way to do this is to use the Fishbone diagram of Man, Machine, Material, Method, Measurement, and Mother Nature (6Ms). Fishbone is a great tool to help a person bucket their information by only thinking about the contributing factors related to each category. Listing all potential causes under the man category, then method, and so on will support creative thinking related to each bucket. All ideas about potential causes should be heard, we want to be open-minded to what is possible. This is about driving better root cause analysis. By building this Fishbone, now you can ask do any of these possible causes contribute to the root cause? This is an area where lots of people struggle. The Fishbone is about finding any possible causes, but many people are trying to find the root cause when listing out the possible cause in the Fishbone. We want to take the possible causes and drive them down to a root cause with the 5 Whys. So, think of general possible causes when using the fishbone. For example: maybe we state, setup of equipment as a possible cause. By using this thinking of general ideas for possible causes, it allows you to go in more than one direction when you break it down with the 5 Why process. Once you go through the 5 Why process, you will have driven down to what you believe to be the root cause. This root cause will be based on your hypothesis which is derived from your experience and current understanding of the world at that time. In our example, we listed out many possible causes. After going and seeing, we learned the vacuum brick on the hoist was worn out and needed to be replaced. When we did our 5 Whys, we determined that there was no preventive maintenance process for changing it out before it started to fail.

In step five, we need to come up with some solutions and decide which ones are realistic. This is also a time you would need to get leadership buy-in because you may need to spend money or change a process. Remember the PDCA thinking I was telling you about, well this is still the planning part of this thinking process. Another question we should be asking ourselves at this time is does our solution prevent the problem from coming back or is it just the correction needed to get back to normal? We always need to put a process in place to ensure the problem does not reoccur. In the example, we listed out the solutions and weighed them for what we felt was most important.

You may want to try experiments to see if our solutions can solve the problem. By testing our ideas, we can quickly see if the solution will work. If there is more than one solution approach, they may both need to be implemented, but on a staggered timeline to validate each solution is working as intended. For example: if we say that not all setups are done correctly because we have no standardized work and people do them differently. We can test the new standardized process and see an improvement. This should be simple and fast. This is also when we start to move from the P or plan phase to the D or do phase of the PDCA cycle.

Next in step six, you need to develop a plan to make the solutions happen. This plan needs due dates, who is going to do the task, and the status. The leader of the A3 will use this action plan to make sure all the important details get completed on time. If we fall behind, we would get our leadership involved to help remove the roadblocks. Additionally, a MUST DO part of this plan that many companies fail to do or fail to understand the importance of is documenting the process changes or the creation of new standardized work. This sets the stage to kill the problem forever and if you don't develop a standard or a process, you will be scratching your head and asking how could this happen again.

In our next step (step seven), we will want to track your success for 90 days after the plan is complete to ensure all your team's hard work has continued to sustain the gains. The team should set up a regular cadence for all to meet and discuss the performance to plan and any abnormalities that have occurred. This is the C or checking part of this process and PDCA cycle.

Next in step eight, you may need to act on your learnings if the tracking is not showing the desired results. If you are achieving the desired results, you may look to identify where this solution is applicable elsewhere in the production system. This, of course, is the act part of the PDCA process.

The key to this thinking is that we do this again and again. You have a problem, you correct it and make the metric better, create a standard, monitor it, and repeat.

Strategy A3

Strategy and making it happen can be very interesting. How do we tell our plan and get everyone on the same page? It can be done using A3 thinking. Policy deployment uses A3 thinking also. Strategy A3 is about telling our story for all to understand the thinking process and how we plan to win. It will also track our success and what we have learned on one sheet of paper. We have always seen the first three steps to a strategic A3 as the charter, the rest are the thinking process and project management. So, let's break down strategy A3 (refer to Figure 2.2).

Step one, we want to make it clear to all why this needs to be done. What is our reason for action? This should always link it to a Key Point Indicator (KPI) metric. If we cannot do so then we need to ask if this idea is something that our company is asking us to drive. Once we can link it to the KPI metric, the problem needs to be clearly stated so we understand what we are facing and wanting to improve. The information needs to be simple and concise, so everyone understands the situation. Using a graph or picture is often times the best approach to help explain the problem. The scope needs to be considered at this time to prevent creeping out of bounds for the strategy. This will help the team understand where they can make the best impact and changes.

Next, in step two, we want to explain the current situation. What data do we have to support the problem or project we are trying to change or drive? We can use our current metrics that tie to the reason for action. Again,

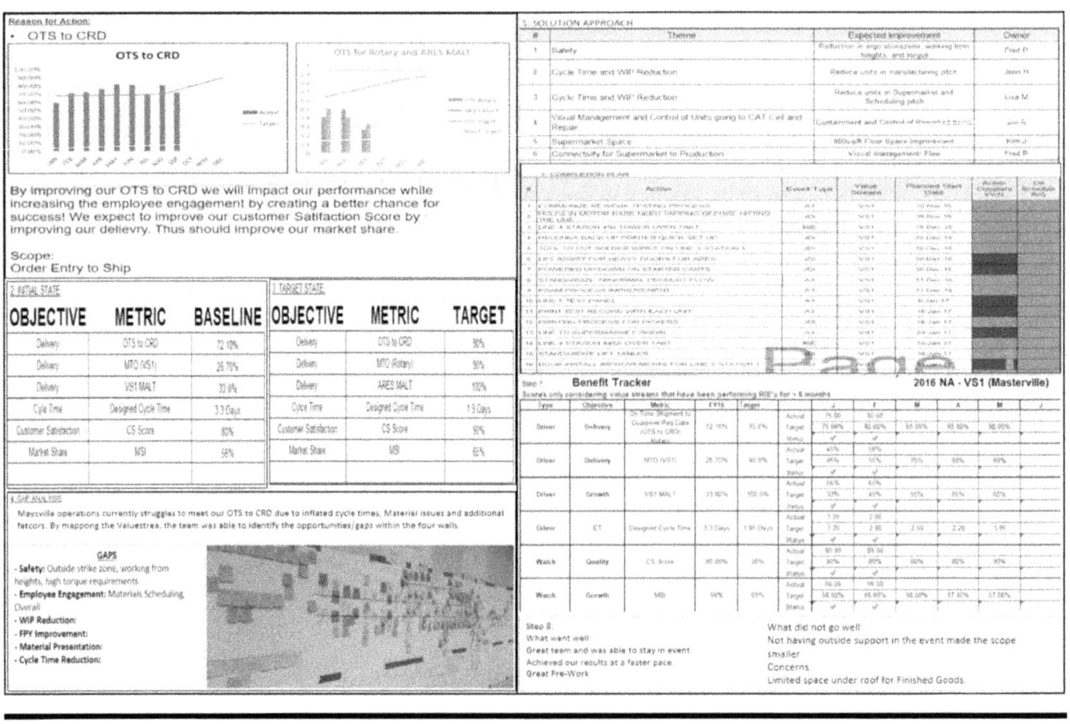

Figure 2.2 Strategy A3.

graphs and pictures help. Fewer words, more visuals. Remember our A3 is about telling our story, our communication venue.

In step three of this process, we tell our people how far we want to go. We need to set the bar. Sometimes we just need to set our goals to meet the new annual operating plan (AOP). For example, we may have been given a goal to reduce our operating cost by 15%. As the company sets the overarching AOP for the company, at the site level this plan should stretch each location. These lofty targets should drive the site leadership to think outside the box or as some might say "drive breakthrough thinking". What is breakthrough thinking? Remember when President Kennedy challenged the American people to go to the moon and back safely by the end of the decade. Well, that challenge is what drives real breakthrough thinking because we had never put a human in space at the time. Breakthrough thinking will take us past our AOP goals, it will take us past any of our customer goals. It is about winning beyond what anyone expected.

Step four is the analysis section of the A3. This is where we break down the information to find solution approaches that support the direction we are driving the company or site. This is still telling a story remember, so again people will need to be able to see the logic in the conclusion to buy in to the direction. Did we use a brainstorming meeting to get here or was it a workshop? Did we use customer interviews, research from the internet, or an impact matrix? Either way, when we use those types of additional information during the process, it is important to make it clear that those resources were a part of the conversation. If we were running a value stream exercise, we would show a picture of our map and the tools we used to complete it. We may have used a reverse fishbone to show what our cause will affect. There are many ways to brainstorm with your team. We just need to explain how we got to this point. In our example, we used a value stream map workshop to understand where our opportunities were most visible.

In step five, just as we saw in the problem solving A3, we want to document our solution approach. If possible, we want to calculate the expected outcome to define what percentage each solution will close of the gap, and plan when the team expects them to be complete. This will give not only the site leadership but the company leadership line of sight as to how the targets will be met as the year progresses.

Step six will be your detailed action plans to drive to completion. These plans must not only list all actions needed but an owner and a defined timeline for accountability. This is like a project planner to manage the team and support when the actual benefits can be realized. Most of the time this is a 90-day outlook because most strategy A3s are yearlong events. In most cases, you will have a parent A3 and a child or children A3s to manage this for the year. The reason for this is because it may be too much information to put on a single A3. The A3 is about communication and tell our story. The plan should be reviewed by the A3 owner and team once a week to manage the program.

Step seven is for tracking your plan. Are we winning or not? Most people will use a bowler chart or scorecard. In this chart, we would want to list the KPI metrics we are trying to drive with the jump-off point from where we started and the goal we are trying to reach. This should be reviewed monthly by the leader to support in removing roadblocks. If we are not meeting our goals, we will need to start a problem-solving A3 to get us back on track.

Step eight is about the team's reflection over what they learned from this A3 and how they may need to act upon what was learned. We want to document the good, the bad, what went well, and what didn't go so well. Spending time reflecting back on the lessons learned in an after-action meeting is how the team will grow and can improve our planning from all that we saw for the next A3.

Proposal A3

The Proposal A3 is similar in structure to the other A3s. It is about what we would like to do and how we would manage the idea. These are most often used to build a business justification case for buying a piece of equipment or possibly an engineering change that would require additional resources. Just think about how often we buy equipment and it is not what we need. (*Note*: I recently worked for a company that spent roughly one billion dollars on several pieces of equipment, only to remove it just a few years later because it did not accommodate the need.) Maybe we have a vision for the future and we are buying something that does not align with that vision. As a site leader and key player in the decision to support the capital expenditure, you may not fully understand the technical aspects of a machine but know and understand the vision the company is aligned to. This vision can get lost in PowerPoint presentations during a lengthy slideshow on how we need a piece of equipment and how it will put us in a better place to serve our customers. There should be many questions to be thought through to answer not only the current state of the business but how this decision impacts the future state of the business. That is where an A3 can clearly show on one piece of paper the business case for the decision. So, let's break down the proposal A3 (refer to Figure 2.3).

Step one is about why we need to move in a given direction and what problem we are facing. Let's say, we have an aging machine that is not performing, costing us money in downtime and material. This is where we would give that explanation and build our business case. If the company has future goals or direction, that information needs to be included here so we keep it in our line of sight. For example, the future state may be to buy all new equipment over a period of time, or limit to specific brands of equipment that meet the needs of production. Just remember to explain and show how everything aligns with the company strategy.

Step two is the same as the strategy A3. Explain the current conditions in detail. What is this costing us now? Labor, parts, expediting goods, over time, and material cost. We want to show the detriment of what the current situation is costing the company. Keep it simple, graphs and pictures can help tell the story.

Step three is the expected improvement for making this purchase. Quantify how the improvement will close the gap related to the current

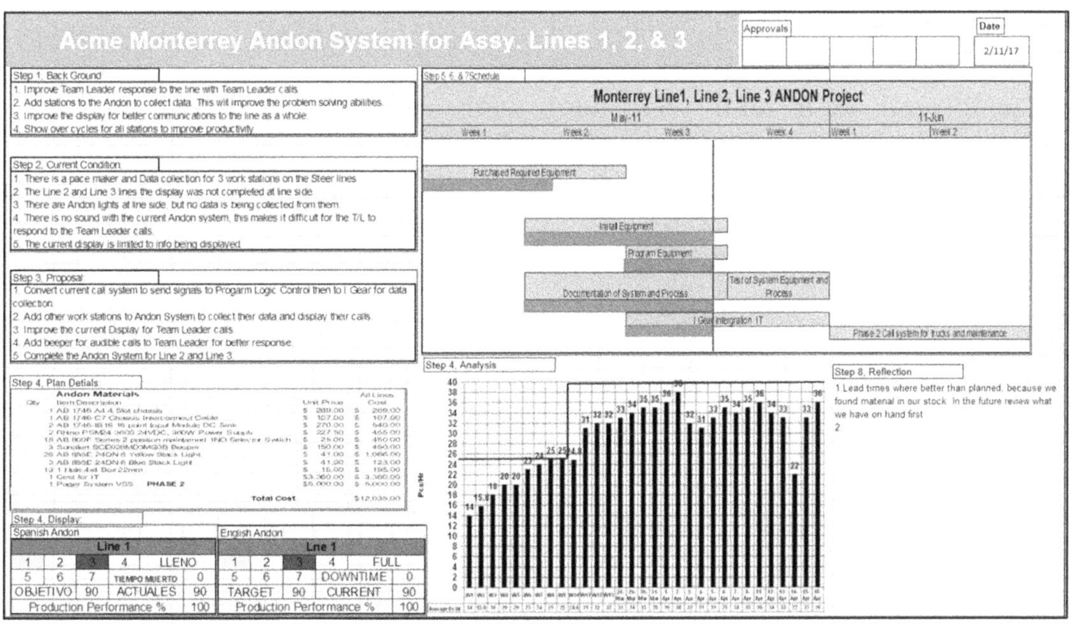

Figure 2.3 Proposal A3.

condition. We may also show the alignment for future strategy and the payback period for the equipment.

Step four would be your planning for what is needed to be successful in buying and installing this equipment. The analysis of your team's thinking to make it happen. Possibly list out part list and cost. Drawings of the future layout of equipment, material flow in and out of the area, lighting needed, electrical, and anything that would apply to the use of the equipment.

Step five would be the overall areas that need to be covered. Installation of the equipment, purchasing of equipment, training to run equipment, materials for routing or schedule changes, and any other stakeholders that would need to be part of supporting the planning for installation.

Step six is the detailed plans to make each of these items happen. There should be a weekly meeting to manage the progress and to plan portion of the process to ensure all tasks will be completed.

Step seven should be aligned to step three to confirm if we hit our target. You want to track the expected outcome. If you stated the change over time would be no more than 10 minutes, track it to make sure we got what we expected. If we did not get what we expected, then a problem-solving A3s may be involved to correct this metric. In our example, we combined steps five, six, and seven with a Gantt chart.

Step eight would be lessons learned from the event and what we can act upon. Identifying a list, the delays, or other issues, and how we can correct them next time. Anything that slowed us down or was not as planned should be reviewed.

As you can see in all the A3s, the first five steps are about **P**lanning and telling our story. Step six is about **D**oing. Step seven is about **C**hecking our progress. And step eight is about **A**cting and learning from the process. So, the Plan, Do, Check, and Act are applied through all A3s. The planning portion is the most important because the more we plan the faster we can do and the better we can win and learn.

A3 Coaching for Problem Solving

There are several things to be looking for when coaching or reviewing an A3. Ask if this was a team sport or if they did it by themselves? A3s are meant to be a group activity most of the time. It involves brainstorming from a team and buy-in from others to ensure success. Next, never skip steps in the process. Each step has key information that connects to the proceeding step to drive the correction of a problem. In step two for the problem-solving A3, there should always be a POO and a PP. The POO and PP will be used in step three and the fishbone in step four in the problem-solving A3. You want to make sure each step clearly connects each step of the story being told. Step seven (measurement) should always equal step three (target). You need to measure what you expected to change. This is how we know when we are winning. All the steps should link together and flow clearly. It's not about how pretty the document is or how it is being filled with graphs, I don't look for spelling or word combinations. I have always seen this document as their artwork. When coaching the owner, don't beat down the work, you want to ask thought-provoking questions to generate a conversation that adds to the owner's learning and brings about an "AH-HA" moment. The A3 process should be about the development of your people as well as managing each problem to a sustainable solution.

Always remember the Lean philosophies. How does the A3 help our customers? Who is the customer in the A3? Show respect to our people when coaching the A3, use the A3 process to build skill and capability in your team. Did they go and see when working the A3? Ask about the go and see and what they learned. Last but not least is the continuous improvement part of using the A3 process. How can they spread what they learned from the A3? Are there other areas that can benefit from the last problem solved? Do they have standardized work to ensure the sustainment of the improvement? These are just a few tips to help you be a better coach of the A3 process. Learning by doing will help you even more.

One more thing about A3s and A3 thinking. You may have noticed the different formats of A3s, there is a reason for that. We want to send a message to not let the document get in the way, in terms of the structure. It is about the process and telling the story. There are people out there that believe we should only handwrite them. We agree to a point. It makes the document and process personal. Using a pencil allows you to erase and correct as you learn or gain new information about the problem you are trying to solve. When we fully understand this, we should be able to use a blank A3 size paper and complete our A3. Even using an excel document you often have to change the size of the boxes as needed. Tell your story so all can understand and learn from what you have done.

Chapter 3

Creative Thinking

Why do we fail at coming up with better ideas? Why is our competitor passing us up? How do we win in every area of our business? Companies often miss the mark when it comes to building a culture of creative thinkers and problem solvers. Largely this is due to an inherited culture within the leadership of a company that stifles or completely shuts down creativity and innovation within their workforce. I recently had a conversation with a Senior Manager, and he stated that he "did not need his Team Members, Team Leaders, and First Line managers working on making things better or problem-solving. I just need them focused on building the product, I can fix all their problems in about 10 minutes". So, my question to this manager was simple; "What are you doing to leave these people better than how you found them? Didn't you just say that every time you get moved to a new area the one you left falls apart"? What this leader failed to realize is that he is shutting down his people's ability to think and in essence stifling or killing his people's creativity to improve or solve problems and in fact not leading them at all. A true leader elevates his charges. More time than not creativity and kaizen come out of the sheer need for change. Creativity starts with outside-of-the-box thinking. People often tend to limit themselves to only thinking about what we know we can do, or what we deem possible. Examples of this mindset, a fixed mindset would look something like this: I am limited to only doing what we have always done. I am scared to try something new because it may fail, or there is no budget to be able to do it right so why bother at all. These are all creative-thinking killers, don't be a victim to the past. I have been on projects with multi-million-dollar budgets, to projects that I had to go behind the back of the plant and dig through

DOI: 10.4324/9781003194781-3

the scrap pile for resources, aka dumpster diving. Companies typically have a "boneyard" so to speak, of leftover material, conveyors, old flow racks, lift tables, etc. that can often be a valuable resource to a creative mind. When working in a Lean environment, this is the place we should always go first to get supplies.

The Team

Teaming and team building drives creativity. Now we just need to put all the supplies that we have out back together for the new way of doing a process. Utilizing maintenance is a good place to start, we would also look at the team members to build our teams (refer to Figure 3.1). Generally speaking, your team members have more talent than just making the product. You may find that many have backgrounds in fabrication, mechanics, maintenance, and much more. If you are going to build a team for *kaizen*, we should be thinking about building this team with people that have diverse skill sets. The best teams have expertise in a variety of backgrounds. Building a diverse team sets the stage for not only a high functioning team but one that

Figure 3.1 Kaizen Team.

can grow and benefit from each other's strengths. A diverse team will have a refined perspective that allows them to gain buy-in from their customer when implementing an idea. They will get the person/customer they are impacting to be involved in the thinking process. They become part of the solution and they will help make the implementation work.

One driver for these teams should be the challenge of paying for themselves. Not literally speaking, but how their work helps the company. This will prompt the team to think of ways to improve the business, that out-of-the-box creativity that companies need to generate a competitive edge. A good leader will make sure they are building their teams by encouraging them in their creative thinking. You will find that these teams will often take ownership of certain projects that would normally be contracted to outside companies, saving tens to hundreds of thousands of dollars a year. As I mentioned paying for themselves!

This team should also be viewed as the next leaders in your company. If you have a high-performing team member, this may be a great opportunity to develop them into your next team leader and so on. The team should be developed in Lean principles, leadership training, and exposure to opportunities that stretch their growth. This training ground for future leaders is an added benefit. This is how I got my start; I was an operator on the line, then moved to a project team, was given some additional training, exposed to working with others at many levels, learned how to get buy-in, put in charge of small projects, then worked my way up to larger and oftentimes multiple projects. These large projects included strategy development, managing a budget, planning schedules. I got some of the most advanced training from the company, this training was on-job training (OJT) with true coaching for performance. I was one of many folks on the line building the product, doing the same thing over and over and now I was part of a highly trained team that can take on any project that was put in front of us. The team had a resume that most managers could not compare to. By doing this, the company had developed their next leaders, we had been developed, vetted and were being promoted all the time.

Out of the Box

Now that we've explained how a good team can be an advantage to your company, let's step back and talk about out-of-the-box thinking. Simple is always better. I have seen complex systems put in place to very basic ideas being used. I have always thought that simple is best whether it was high cost or low cost. I was coached during my creative thinking development that I should always work to apply these four words: Safe, Simple, Reliable, and Cheap. I have had leaders that thought that simple was to be without pneumatics and electrical controls. I would say we should always think this way first, but don't confuse this as the only way to think. Simple should be applicable to these components: simple to use, simple to repair, and simple to change. Obviously, safety would apply to everything we do and should be our first priority. Ergonomics would be taught to this team; this way they are aligned to the environmental health and safety (EHS) requirements and can educate the operators they engage during project implementation. If it is not safe, simple, and reliable people will not use it and they will default back to their old process and they will lose confidence in the team.

All ideas should be trialed in a controlled environment, creating a potential hazard to our people or any type of impact on production is unacceptable. I also strongly believe in layout drawings, marking up the floor so everyone can visualize the change and how it is going to affect them and the process. Although I did not use the 3P method (Production, Preparation, Process), companies may desire their teams to use this methodology to display the changes.

The team should be thinking of speed when it comes to change. Everything should be done in a short period of time to show that change can happen fast. If we serve our people in a fast manner, it can show that we care about making their processes better now. This will create synergy between the operators and the team making the changes, setting the stage for people to be an advocate for change. One thought to keep in mind during these changes is a word I like to use "delighters". Delighters are the little things that may not be asked for, or they may be asked for but are out of the scope. Delighters are things that make life better for our person dealing with the change. One example in our daily lives may be something such as heated seats in a car in the past were a luxury, now many car companies are making it a standard versus an option. That is a delighter. I see delighters in the process as ideas that take the mental burden away, that make the process physically easier, for instance, a water bottle holder added to the workstation. This is about changing our people's perspective and help them to be less stressed and tired each day. It makes training a new person easier, which helps the supervisor's life. This also makes your new team the "go-to" team for new ideas to be created. The continuation of building more and more people that support process improvement through creativity.

Your kaizen shop should be a living example of Lean. Visual management, 5s, kanban, reduced movement, efficient, safe, flexible, and always changing as we learn. Think about it, if we have a shop that is never clean and you cannot find what you need, then you will be slower at making a change on the production floor. Lead by example in the shop and the office.

Should this team just be for fabrication and making cool innovative creations for the production floor? The answer is no they should be trained to be like an industrial engineer. They should be able to do time studies, write standard work, help with model launches, project leaders, and other opportunities to develop them. A good leader needs to develop others to be good leaders. If you are a good leader then you are an example for all others.

You now have your team, you are developing the next-level leaders for your company, and people will see this and want to be part of it. People want to be part of something bigger then what they are currently doing, and this kaizen team will be viewed as a good thing for your people and the company. Before you know it, you will see people thinking out of the box, and thus driving a creative culture in the workforce. This culture is in my opinion the "secret sauce" that Toyota has and everyone else is chasing.

Understanding the 3Ms "Muda, Muri, and Mura" (Japanese terms for three types of waste) or the 8 waste that often exist in our processes is a critical part of a Lean journey. The acronyms DOWN TIME or TIM WOODS are great names to help people remember what each waste is. In this example, we will use DOWN TIME.

Defects (bad quality)
Overproduction (producing more than is needed)
Waiting (for information, material, the next process, etc.)
Non-Utilization of Talent (this would be support functions, our team of talented people, etc.)
Transportation (movement of material)
Inventory (in excess, more than we need, cashflow)
Motion (movement of the human body or a machine)
Excessive Processing (doing more than is needed to complete the process)

Defect would be the D in using the DOWN TIME theme. Defects create rework, loss of material, extra movement, and worst of all hurt our customer and their confidence in our ability. So, we need to focus on building quality in each process. How can we build quality into our daily processes? The first step would be to have standardized work that we repeat every time. That is the foundation to our success of a repeatable process and continuous improvement. We can also put Poka-yoke in place. What is Poka-yoke? It is error-proofing a process. It could be a fixture that only allows the part to fit in the correct way due to a structural feature on the part and/or fixture. It could be part of a program to detect a defect or the design of the part so it can only be assembled one way. We can audit the process but this would be after the defect was made. So, auditing should only be used for processes that we have not found a way to Poka-yoke. Auditing is a waste and the customer does not pay for this.

Overproduction is when we produce more than is needed at the time or pace for our customers. We often do this for a variety of reasons such as a lack of good scheduling or poor changeover times. For whatever reason, we need to understand that this is by far the worst waste we can have. Think about it, if we overproduce, we could have defects in our parts that are hidden. This would drive more rework than we would have had if the amount was less. We would have extra movement of the material and storage. It takes away from our capacity of equipment and hides the opportunities we should be working on. It would force us to have more raw materials. Material that could age or get lost. It can stress the supply chain based on forecast expectations. In the end, this waste has the ability to create all the other waste and will create these in a struggling production environment, which perpetuates the cycle.

Waiting can be easy to see. Most of the time we find that our people fill this wait time with busy work. Why is that? There are two possible paths for a person to take. One path is related to "if the boss sees me standing around, they will give me more work to do". The second is, they want to win and do the right thing for the company. So, when we don't remove waste out of the process, we may not give our people the respect they deserve for the time they are away from their families. In my experience, I have had several conversations about waiting and the need to build some waiting into a process for the reason of fatigue. So, let's talk about fatigue, people need to understand that first, we have breaks during our day and additionally, we must only build our processes to 95% of the takt time or pace of the process. The 95% allows for people to take a drink in between units and if we rotate our people through different processes this will also reduce fatigue. Additionally, reviewing the ergonomics of the process will improve the safety and fatigue factor. In the end, waiting must be driven out of each process to be an efficient competitor in the industry.

Non-Utilization of Talent is what we were talking about in the opening of this chapter. We have very talented people in our workforce and support functions that need to be identified for continued growth and development. These folks need to be drawn into either support in the manner of their given support function or simply to be leverage for whatever the best skills are. Whether it's as a welder or a computer whiz, seeking out and utilizing these skills is often overlooked. Leveraging those skills around you follows the Lean philosophy of respect for people.

Transportation is the movement of materials and information. The movement of material and information does not change the shape or the form of the product so it should be seen as pure waste. If you were to build an ideal state for your processes or value stream, you would want this removed or reduced to almost nothing. Our customer does not want to pay for the transport of material or information, so why should we. Understanding the flow through your manufacturing site and moving processes closer, understanding scheduling, reducing changeover time can all contribute to this improvement.

Inventory – you need it but, how much? It hides problems that need to be improved. Inventory is cash flow that could be used to improve the business. Think about it as if it was your company. Let's take money out of your wallet and make a product that we don't need at the current time, so that we may sell it later if it doesn't get damaged while waiting to be sold. Then we must go to the bank and get more money to improve our process and now we pay interest on a loan. Get the point? Inventory is our cash.

Understanding the correct levels of inventory is critical, the minimum level versus the maximum level. Often referred to as simply min/max levels. These min/max levels should trigger either a re-order or a stop I'm full notification. Identifying these levels take some time. Max, well most companies have too much inventory for fear of a shortage. While min is not always a calculation, one needs to continue to drop the re-order number by one until it creates some pain of a shortage, then you would understand the minimum number of parts you need to run the business. During this time of testing, it is wise to have a strategic stock set aside to avoid the actual pain of shorting the production system. We only want to stress the system during the test, not affect production.

Motion, next-level thinking about wasted movement is a key concept to be taught to this team. How do we reduce motion in our processes or reduce motion in a project? Motion is a waste; we should be observing and reviewing a way for it to be removed. I often ask people how they define the difference between motion and transportation and rarely receive the response I'm looking for. The way I differentiate the two is that transportation is the carrying of any material or information. While motion is the movement inside of a process. My go-to example is back when I was in the automotive industry, I had an operator that was having trouble keeping up with his process. Every five or six units, he would need help to get back into his work pitch. As I stood there and observed him installing some lights in a trunk lid, I began to see the waste of motion causing him to be behind. He was installing two lights that required three nut shots each, as he would shoot each nut, he would bring both hands down to his waist, load one nut, raise both hands back up to his head level to make a shot, and then back to his waist each time. Basically, the operator was about three to four seconds slow on each unit which leads to needing help every five or six units. I engaged the operator in some conversation about moving his arms up and down for each shot and asked him to work with me by shooting all three shots before bringing his arms back down. Before we knew it, he had changed the way he performed the process and had gained three extra seconds. In short, by not bringing his hands down for the six shots, he went from three seconds per unit behind to three seconds per unit ahead! The waste of motion was gone and sustained by writing this method into the standard work document. My point is this waste is not easily seen by the untrained eye and the development of a critical eye is a key to success!

Excessive processing is when we do more than needed to make our part or to complete our process. This takes time away from our available capacity during the day. The customer doesn't want to pay for this either and we need to look for it and reduce it. I have seen people over or excessive processing and they don't even realize that they are doing it. I remember running a monthly report that was taking me an hour to compile the data and submit it. This was how the process was designed and I was following the standardized work. After some thought and improvement to the spreadsheet, I was able to add a macro that reduce the hour-long process to ten minutes and improved the accuracy of the document. How many times do we report out to our leaders the same data in four different ways? There are many forms of how we do excessive processing in our daily work. It's our job to find it and reduce it.

My manufacturing example is pretty basic, but I see people doing excessive processing all the time. While I was working in the air compressor industry, I had an operator pre-loading their bolts with a lock washer and a flat washer (approximately 50 bolts). The washers were required to build the compressor and the operator was convinced that they must pre-load in between each unit or they could not be successful. This is a common behavior that is misunderstood and mistaken as needed work to be done.

Muri is overburden of a process. This is the stress that we add to poor equipment performance, mental burden, and the overall production system. This is the poorly defined process that makes it difficult to find or do a process. This can be the physical stress of not having the proper equipment to assist in moving material. This can be the mental stress that we have in doing our processes from having poor instructions and an influx of decisions to make inside the process. The mental stress and rework from poor communications. Muri can drive poor motivation, overtime, and disgruntled people. It can also contribute to defects and the loss of capacity in a process.

Mura is the imbalance of our processes. Think about a line of processes with team members standing around waiting on another process to feed them. That is the waste we talk about called waiting. This could drive our people to fill this time with busy work, which over time will feel like necessary work. If we balance the work and fill the available time up with value-added work that the customer is willing to pay for, we find out we have more people than needed to get the job done. We can then move them into a role that might be missing a person to do it.

Muda, Muri, and Mura all require creative juices to drive innovative thinking to solve issues within our companies. Using a critical eye to identify, visualize, and then eliminate each of these wastes is what gives a company a comp!

Back-Office Ideas!!!!!!!

What about the back-office areas or yearly events? Can we improve the flow in the office? In today's age, do we need a desk phone for everyone, when most of us have company cell phones? Are the printers in the best location for high-volume users? How is the information flow in the office environment? Do we send or receive hundreds of emails that are just to keep us in the loop? Emails and meetings are a real-time killer. I have worked with people that get 300 plus emails a day. The average workday is only 430 minutes minus breaks and lunch, this means ¾ of that person's day was consumed by these emails. That does not leave a lot of time to do anything else but just get through the day. Some might say "most are just paper trail emails I don't need to review, but they do distract me". Next, there are meetings to plan, people to invite, and preparation to be done. All in the name of running the business! I typically work to have the meeting before the meeting. Meaning I host a short one on one with key people to gain alignment and steering the meeting. I will decline a meeting if I believe it is non-valued to me and forward if necessary for someone on my team to attend. Sometimes we just need to review a report and not attend to make our time more useful. Why not make information visible on a board for all to see the latest data? Are the teams close enough that they can support and work together or are they in different parts of the building? Is your office area open or does everyone have their own office? To prevent people from working in silos, we need to get them into an open area with fewer walls between them. These are a few areas to think about in the office. One of the companies I worked at fell in line with the open area concept by not having cubicles everywhere. One key learning is it created a challenge when multiple people were on the phone, so through some creative thinking the company built several "phone booths". Small two and three people conference rooms to host conversations or conference calls. This allowed people in the office to have quick face-to-face communication while providing a location to retreat to when needed. One might say that was "out-of-the-box" thinking. This can affect the morale of your people. This is how we make the office and the grounds of the factory more inviting and how we make it a place that we would be proud to bring our family to? These are the delighters to make your customer, the employee, happy.

Now let's think about our daily, monthly, and yearly events. How long does it take to get a quote for a customer? Could we win more business if we were faster? How long does it take to receive cash from our customers? The monthly close of the books. Can we reduce this time and still get good results? Items like the annual operation plan (AOP) can be a lengthy process of back-and-forth catch-ball process or launch of new products should drive creative thinking to reduce time and the cost to launch. If we measure this and reflect on the learning from the event, we can challenge the team to be creative in finding ways to reduce steps in our processes. Physical inventory is one of those events that challenge many companies and cost companies a lot of time and money. During my time in the compressor industry, one location shut down for up to three days to collect the inventory data. At

Toyota, I remember they would do this in a matter of hours not days. I remember thinking and asking what's the difference? This is one of those instances where Toyota applied some synergy and creative thinking to reduce the pain of the physical inventory process. Many companies spend overtime building up finished goods, so they don't shut their customers down and stop all production for several days to count parts. Some out-of-the-box thinking can help reduce this time.

When thinking about our offices, Lean principles do apply. Mapping out the current process and identifying opportunities to improve all functions of the business! It boils down to "How can we deliver a better product at less cost"? Don't always expect a home run. Look for the incremental change. Everything we do has a cost, time, money, frustration, and so much more.

Enterprise (VSM)

A value stream map (VSM) can be a strong tool to drive creativity! In the industry, using a VSM to find areas to improve should be utilized as direct support to running the business (refer to Figure 3.2). VSM at the enterprise level helps a lot of leaders make the right decisions for all levels of the company. These maps drive good conversations with lots of questions to be discussed. While many people during these events are very focused on all the reasons "why we can't, you can't, they can't", the real question needs to be written on a board for all to see "HOW CAN WE".

While you may find most of your issues in the manufacturing area, we still need to understand the entire value stream and the time it takes for the product to flow through the process. What if we found out that our quoting process was taking a week or more and our competitors were completing the quoting process in a matter of a few days? Does this give them the advantage to gain more business? Most would say yes it does. I'd say that most of us have chosen a product based on fast availability. Suppose we found out it was taking us more than 60 days to receive cash for our products, this should prompt the question "is that a reasonable timeline?" Would that be a project to apply some effort to? In terms of launching a new product, if we take several months longer than our competitor and have many issues during this process, would we find ourselves behind the curve? Yes, a VSM will help to visualize the long lead times in the production system. This tool gives a company or site focus and a plan to go attack issues.

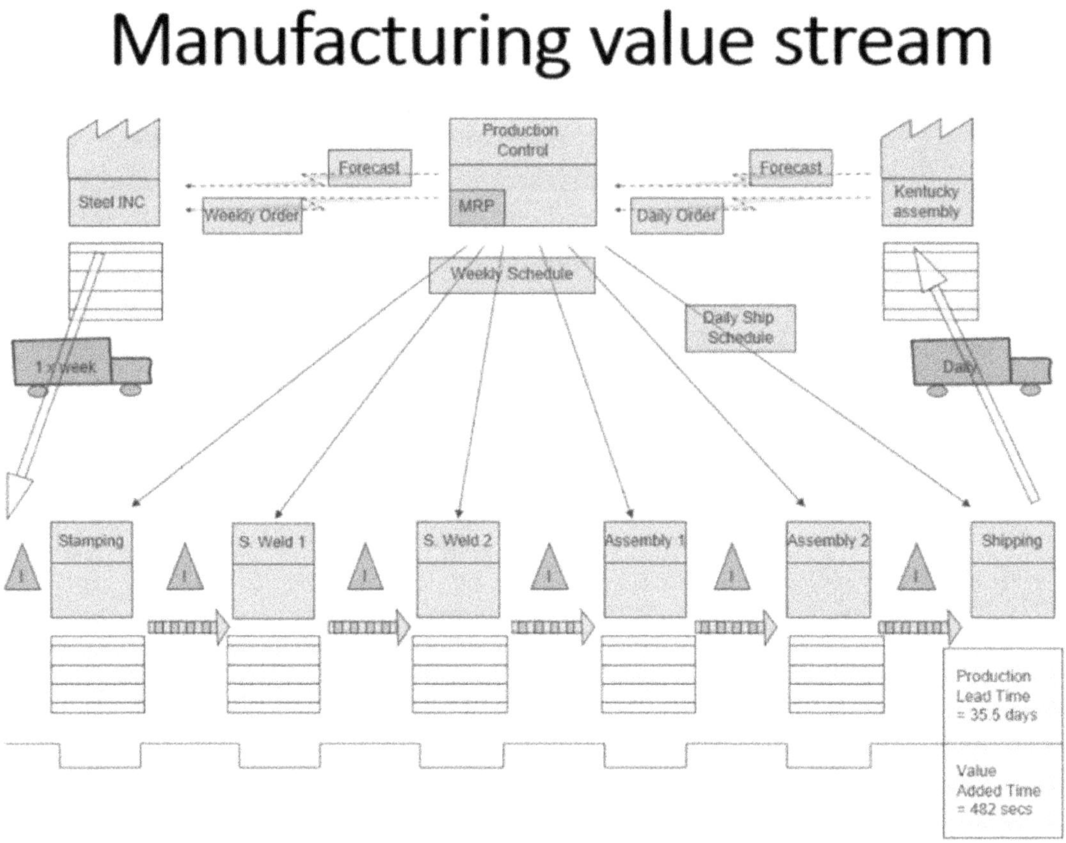

Figure 3.2 Value Stream Map.

When looking at doing these mapping events, we suggest breaking them up into areas. Run the events with the key leadership from each area. Use an A3 to gain alignment and support from the leadership team. For example, manufacturing may not have much to do with receiving cash, so accounting would be the lead. In this case, the leadership from this department would be the main players in the event. You may need support from other areas, if they are in the scope of the planning, they should be invited. Once you develop your plan, it could be a year-long process to complete and it needs to be visible to all. Your team should meet weekly about the progress and report out monthly to leadership, so roadblocks can be removed, and all can be accountable for the results.

You will find that the four Lean philosophies are tied to VSM. First, the VSM will address many customer concerns by providing the end customer a high-quality product in the shortest lead time. The shareholders now have a more cost-effective process with less waste and inventory and the employees have better processes and the ability to work for a company that is passing the competitor. With the use of a VSM, there will be go and see activity as you develop it. This is a very important part to understand what is happening in our processes. It shows respect for our people because we are driving improvements to their processes to make us a stronger company. Continuous improvement will become a by-product of the event. The point about a VSM is not to limit it to just manufacturing, it can touch many areas of the company. As we begin to understand what our customers want, we are also understanding how we are competing against our competition. If we can over-deliver the expectations of our customers, where would that put us in the market? If we understand how our competitors are performing and we are beating that, where would that put us for growth?

Supply Chain Support

This is an area that can be overlooked. If we think of the customer first, we would/should make sure that the supply chain components that our finished products consist of are stable, consistent, at a good price point, and of great quality. Most companies focus on the price point by negotiating only, without any support to get to a better price. The quality with consistent stability portion is usually just a key performance indicator (KPI) that we can hold them to. If the supplier is late, there may be a charge for being late, if their quality is poor, we can charge them back, in the end, this doesn't fix the problem. It is just another form of waste in our process. Look at our suppliers the same way as you do our people, and show respect by offering development. Then start negotiating a price point reduction for the help we give in making their processes better. By helping them to see opportunities, they will deliver more consistently than before, with a higher-quality product. This is very much a symbiotic relationship; their success is your success.

In a past life, we had a supplier that was underperforming and we partnered with them to help correct the problem. First, we laid the ground rules for this help by making it clear what was in it for them and us. On-time delivery had been plaguing us for some time now and it had to be addressed. Initially, this was not about a price cut, that comes in time, but if we achieved our goal, they would save money on late fees and we would save money on our end in no disruptions from late deliveries. This supplier also supported one of our competitors, so a question arose from leadership.

By helping the supplier, how much of this is supporting our competition? This is a fair question, but in the end, we were the ones building the relationship with the supplier, not the competition. Also, what if we don't help them? With this improved relationship, we can continue to help and ask for price reductions, which is a win-win strategy.

Additionally, one of our focuses was to support the supplier in reducing the changeover time on their mainline. This would also develop our supplier in the future to be less dependent on support like this. We would teach as we worked with them. This helped to achieve the on-time deliveries to our plant. The relationship became more of a partnership.

As we work with our suppliers, we need to keep them in mind as a partner in our journey. We should not have the attitude that we are the customer and therefore we are always right. We should be asking; how can we support each other in the success of each other's business? When we find a company like this willing to let us in, we want to make sure you are living the four philosophies of Lean. Always keeping the customer in the forefront, showing respect for each other, go and see to fully understand, and drive continuous improvement daily. This can be a great recipe for success with our suppliers.

Chapter 4

Standardized Work/Standards

Why does Lean fail in our processes? Why does it seem so hard to train people up to make a product without defects? Why do we see production areas not performing well when new people are introduced to the process? Why do we see shift performance so different from shift to shift? Why do we see plants building like products performing differently? Why do leaders say they can't miss work without their team performance dropping?

Standardized Work

Standards are key to many of our issues, or the lack of standardized work is more like it. I have been to many locations and found out they do not have standardized work for their processes. I ask them what do you train with? The response is the operator shows them or they have work instructions. What they do not understand is that it takes longer to train and most of the time we are training them bad habits. Then, most of the time, the operator does not fully understand the details they need to know to do their process every day. You see operators doing the process differently than other operators. We move an operator back to a process they have not done for a long while to find out that the process has changed and they do not know it.

Think about all the problems we can see from this. Defects getting to our customers. Cycle times on the process going over what we expect. Operators getting hurt because they are not following a standard. If people do processes differently, we are compounding our variables for what could be the root cause of an issue. If you make a change to a process to improve it without standardized work, how do you know it can be maintained?

DOI: 10.4324/9781003194781-4

Are we truly setting our operators up for success without clear standardized work/documents/etc.? I recently posed this question to a team I was working with as a rapid problem-solving team related to defect resolution. The situation in reference is, we have mechanics (this is the selected name for people building the product) unloading parts from a protected dolly and loading them onto an elevator floor that has anti-slip paint covering the floor. As the mechanics load these three feet by two feet parts, they are laying the part directly on a portion of the customer-facing surface (the portion in which the customer will look directly at) on the sandpaper-like anti-slip paint. As I watched several different mechanics perform the task, I asked myself, do they understand that placing the part on the rough anti-slip floor could scratch the finish and cause a defect? I wonder who trained these people and by what documents they were trained? How do we the leaders of the company allow such behavior, to handle these parts with such disregard? Or do/did we? How is the standard work written, is it clear to the mechanic what should be happening, or is "tribal" knowledge the way we learn to perform our jobs? Most people call it tribal knowledge, I call it "travel" knowledge because it's the undocumented knowledge that travels away when the person leaves the area or retires. Now back to my original question about the clarity of the standardized work. One of the managers on the team pulled up the document that the company deemed valid to use as a mechanic. Inside this document was written, "Use white gloves and handle all parts with care". Now, we witnessed the mechanics handling parts with care, gently setting them down, although they were still setting them down directly onto a surface that was creating issues. This behavior is directly correlated to what we write into their standardized work in which we train too. If we the leaders allow vague writing, open statements such as "handle parts with care", we introduce subjectivity to understanding, and thus the world of tribal knowledge begins. Companies that want to build a strong system to run the business will/should start with the standards or standardized work that is to be followed by all. The standardized work that I am describing is in this day in age the difference makers of sustainability, profitability, and market share.

Let us define what should be in standardized work. First, we need to understand the takt or pace that our customer wants the product that we make. The takt or pace is set by the demand and the available time we have to make it, simple formula. For example, we have 24 hours in a day, which means we have 1440 minutes in a day. If our operation is 24 hours and we take 30-minute breaks for lunch, two 10-minute breaks, and have a 5-minute start-up meeting, we would have a total of 165 minutes we are not making the product. So, if we take the 165 minutes from the 1440 minutes in a day, we now have 1275 minutes or 76500 seconds of available production time. This calculation can be done in either minutes or seconds depending on what you prefer or based on the speed of your line. If you are or intend to build a unit inside of a minute, you would have to use seconds to calculate your takt. Now, we need the customer's daily demand. This number will need to be a level or average number for the month. It would not make sense to change our pace every day. So, if the demand is 2500 units per month and we only work 20 days a month then the daily demand would be 125 units. Now we have the information for our takt time formula. We would divide the 1275 minutes of **A**vailable time **B**y the 125 units of the **C**ustomer demand, giving us a takt time of 10.2 minutes per unit. It's as simple as ABC, Available time, By, Customer demand. This would be the takt or pace that we should be making our product. Understanding this is foundational for designing the processes to build the product (refer to Figure 4.1).

The second key part of standardized work is the work sequence or order of work elements (refer to Figure 4.2). This is the order you want the process to assemble or to work in making the product. It is important to make this part clear because you want all to follow the process in a repeatable way. If all the people do the process in the same order, we reduce variables that can add complexity to a problem. If each person does a process in a different order, one person could be adding walk time and not realize it. Another could be adding extra handling to the process. While another could be adding a safety risk to the process. All of these together make it harder to get to a root cause of a problem because you would have to watch all

	Shift Breaks		
	1st	2nd	3rd
Lunch	30	30	30
Break 1	10	10	10
Break 2	10	10	10
Start up	5	5	5
Sum	55	55	55
Sum of all breaks (min)	165		
Total Minutes per day	1440		
Available time (minus breaks)	1275		
Available Seconds to work	**76500**		
Demand per day	**1260**		

Takt = 61

$$\frac{\textbf{Available Time (sec./day)}}{\textbf{Customer Demand (pcs./day)}} = \textbf{Takt Time (sec)}\ \frac{\textbf{76500}}{\textbf{1260}} = \textbf{61}$$

Figure 4.1 Takt Time Calculation.

Symbol	Step #	Process	Work Time	Wait	Walk Time
	1	Collect widget	2		
	2	Collect bracket	2		1
✚ ◊	3	Remove welded part from fixture, inspect weld, and set on holder	5		2
	4	Set widget and bracket to weld fixture	3		
✚	6	Collect welded part from holder	2		
	7	Start welder	1		2
	8	Set welded part to transfer rack for OP 30	1		1
	9	Collect sub bracket and 3 plugs	4		1
	10	Set sub bracket to holder and Secure the 3 plugs to sub bracket	10		
◊	11	Collect hose and connect to bracket	8		
	12	Collect sub bracket from holder	2		
	13	Set to transfer rack for OP 40	1		2
	14				3
	15				
	16				
	17				
	18				
	19				
	20				
✚ Safety		◊ Quality Check Work in Progress	41	0	12

Figure 4.2 Work Sequence or Order of Work Elements.

of them start to understand what is going on. We all need to understand that standardized work should be the best way we know how to do a task or process at this current moment. If we believe we have found a better way, we should control the change and document the improvement to gain buy-in from others. Then change the standardized work to the new best way that we know. Another key point to work sequence is – it should be short simple descriptions. The reason for this is – it is a training document. We want to break the process down to make it simple to understand and follow. This will also be part of the job breakdown sheet (JBS). The JBS is a supporting document to the standardized work document. The JBS will cover the work sequence, the key point, and the reasons why you do each step in the process. The JBS can have pictures and a detailed explanation around this portion of the standardized work. We will need to document the cycle time for each step. For clarity, the difference between cycle time vs takt time is, cycle time is the actual time a process or task takes to complete, while takt time is the time a process or task must meet in order to meet customer demand. This is important for problem-solving and training. The overall cycle time should always be less than the takt time. If it is greater, we will not meet the customer demand.

The third key part of standardized work is the work in process (WIP). There should be an area on our standardized work that shows the process from a bird's-eye view. This will show the flow of the process and the furniture or equipment used. It will also show the correct inventory levels in between operations, this would be called WIP. The WIP in this illustration is circled in red. It is very important to establish a standard amount of WIP, this gives us a control for how much is allowed. The flow is shown below with circles with numbers inside. The numbers represent the step or element and where the team member should be standing during this step. These circles are connected with arrows showing the flow of the work pattern. The last part of each process is a return to the beginning to start over. This is shown with an arrow that has a dotted line. You want to make sure the equipment, flow racks, tables, and anything in the work area is a good illustration of how the work would look from an overhead view. This will give us a good picture of the flow and obstacles that are in the process. These items are items important to understand because we want to know what normal should look like and it gives us control over the process (refer to Figure 4.3).

Once you have standardized work for your people, it should be displayed near the work being done. The reason for this is not to have them read it all the time, it is for problem-solving. Think about it, if there is a problem in a process, do we want to waste time looking for our standardized work document when we could already be reviewing it to find the gap that needs to be solved? We need to have an understanding that anytime we see a problem with the performance of our processes then we should be getting our standardized work out to review it. If we do so, we can find the part that is not meeting our designed standards, then we can get to a root cause. Let's give a couple of examples. If we are not meeting takt or rate, we can review the cycle times of each step or element and find the one that may be taking longer. If it was one step or element that is off the pace, we can dig deeper into why and correct it. It could be that the team member is taking too long between a step/element. We could see that the time is off, and also see if the walk pattern is correct in the bird's-eye picture or work area drawing. The main point we need to understand about problem-solving our process is that our standardized work is a detailed document showing the

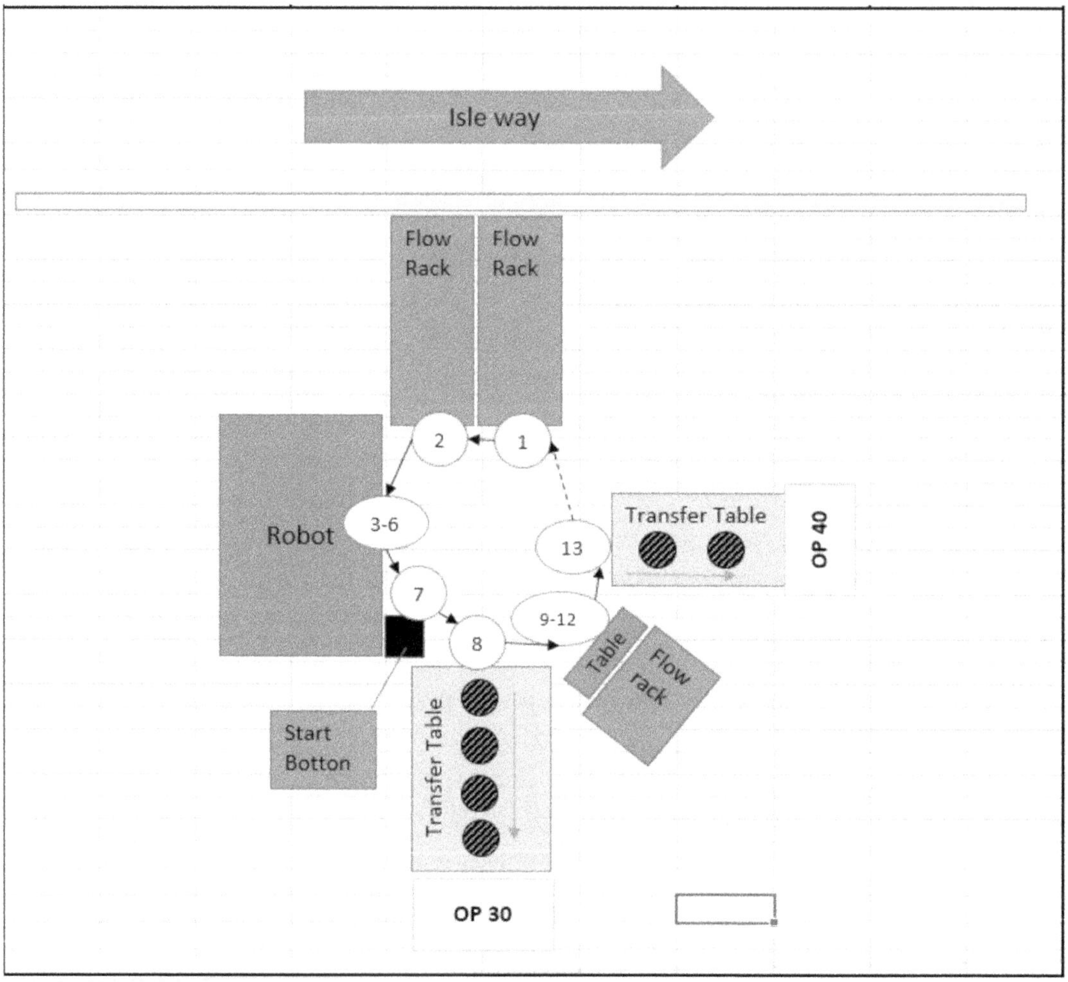

Figure 4.3 Birds Eye view or Work Flow.

order of the work, the time it takes to do each step, the time we expect the overall process to take, the flow of the process in a bird's-eye view, with the standard amount of work in process shown. With all this information, any leader should be able to find the part that is not too standard and begin to bring it back into the standard (refer to Figure 4.4).

By posting it near the operation makes it clear that we have studied the process and this is the way we should be doing the process. The expectation for the hourly performance should be visual and clear to all. Without these items present, how do we know if we are winning or losing in our day? If we are losing, we can quickly investigate why and drive countermeasures. If we are overachieving, we should also ask why. For example, what if the standard is 100 pieces per hour and we are seeing that we can do 120 pieces per hour? Do we have a better standard?

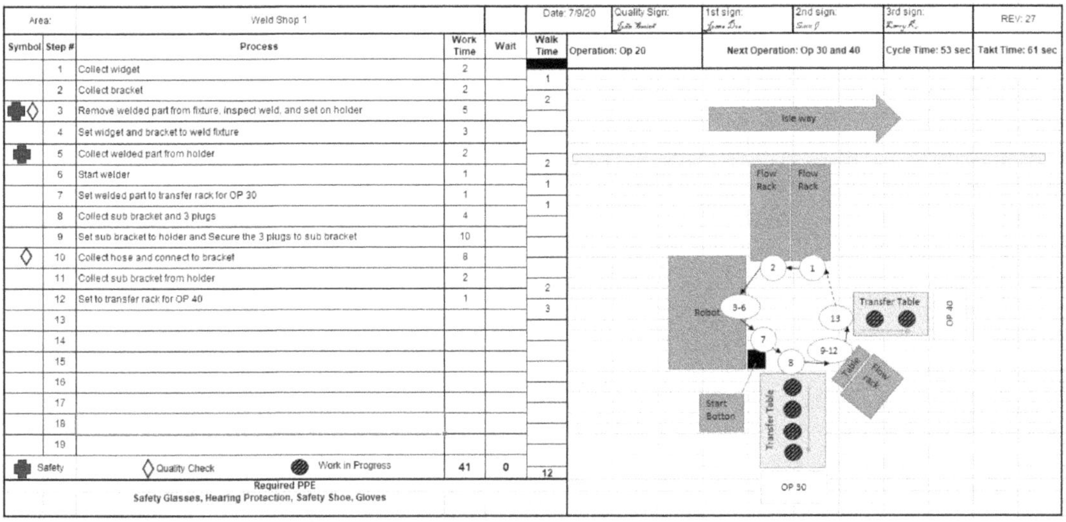

Figure 4.4 Standardized Work Sheet.

How did we get to the better standard that should be our next question? Are we making a good product at this pace? Are we doing it safely? Was the new standard developed in a controlled way? If so, when will we have the new standardized work in place? Always remember we need to protect our customers and respect our people during improvements. We should also celebrate successes when they happen. If the teams and team leaders are driving this, the success is greater and we should build upon this behavior.

Audits are a key point to sustaining our standardized work. The standardized work document should be reviewed to know what is happening at the process. By doing this, you can understand if a process is out of standard and action can begin to get us back to our standard. The team leader and group leaders should perform one audit every day on a different process and different operators if we have multiple operators trained to be able to rotate to different jobs. They should be doing these audits to ensure that we are following our standards and during this time they should be looking for opportunities to improve the process. We always want to be driving continuous improvement in what we do. When it comes to our processes, we want control over the changes we make. This is about protecting our customers and the safety of our people. In the spirit of audits, we should have other leaders involved in audits of our processes. Their pace may not be once a day. The reason for this is to put a layer process audit system in place. By doing this, we can reinforce the importance to our people that standardized work is the key to our success (refer to Figure 4.5).

Job Breakdown Sheet (JBS)

Process Name: Op 20 Part: YZ-785235 Date: 7/29/2020

Dept: Weld Shop 1 Rev. # 27

Symbol	WORK ELEMENTS for WHAT	KEY POINTS on HOW	REASONS for WHY
	1 Collect widget	1.1 Use right hand to collect widget	1.1.1 This allows your left hand to be free to collect next part.
	2 Collect bracket	2.1 Use left hand to collect bracket., then transfer to right hand with widget while walking to robot.	2.1.1 This will allow the left hand to be free to remove part from fixture.
✚ ◇	3 Remove welded part from fixture, inspect weld, and set on holder	3.1 Grap part from top 3.2 Inspect weld for defects	3.1.1 To prevent TM from burning hand from hot weld 3.2.1 To Control defects at process.
	4 Set widget and bracket to weld fixture	4.1 Set parts flat in fixtures 4.2 Set clamps	4.1.1 To insure proper alignment for weld 4.2.1 Robot will not start when clamos are not set
✚	5 Collect welded part from holder	5.1 Collect part from top	5.1.1 To prevent TM from burning hand from hot weld
	6 Start welder	6.1 Double tap start button 6.2 Look for green light	6.1.1 Robot will not start without a double tap on palm button. 6.2.1 Green light is a confirmation of robot starting
	7 Set welded part to transfer rack for OP 30	7.1 Only 4 parts are allowed in stock rack	7.1.1 This is are standard to not over build
	8 Collect sub bracket and 3 plugs	8.1 Use right hand to collect bracket and left hand to collect pugs	8.1.1 Process is set up with torgue tool on the right and allows right hand to be free for tool.
	9 Set sub bracket to holder and Secure the 3 plugs to sub bracket	9.1 Set bracket to datum 9.2 Torgue each plug until beep for tool.	9.1.1 setting to datum allows for proper fit 9.2.1 The beep from the tool indicates full torgue
◇	10 Collect hose and connect to bracket	10.1 Insure hose has ID of blue strip 10.2 Insure fully setted	10.1.1 Bue strip is th proper ID for this part # 10.2.1 If not fully setted hose will blow off in next process.
	11 Collect sub bracket from holder	11.1 Use both hands to collect	11.1.1 Part is heavy, both hands improves safety for handling.
	12 Set to transfer rack for OP 40	12.1 Only 2 parts are allowed in stock rack	12.1.1 This is are standard to not over build

Figure 4.5 Job Breakdown Sheet.

Standards

Standards for visual management are very important for stability and driving improvements. How many times have you seen a new board for metrics or a different board at another site that was supposed to be the same type of info that other sites use? The problem you face is the layout is different and the metrics may not all be the same. You now have to try and adjust to the different boards and what they are trying to say with it. That takes time and adds to the confusion, which could add to the problem of why the board is there. Metrics should be visible so we can tell our story and how we are problem-solving. The problem with different standards from site to site is that leaders from other sites may miss opportunities to understand what is happening so they can add value with coaching or bringing new ideas to other sites. I have even seen this on the same site with different boards, so this is on the micro-level. If you want to run a different style board or metric, it should be a pilot so we all can learn from it, and then we can build upon the improvement. With pilots, we can get buy-in from other leaders and develop good standardized work around it for all to understand.

One might ask how the role of the standards plays in the business model. In terms of sustainability, it boils down to set the stage to define all details of a given role from "leader standard work" to the standard process that every employee must follow. These standards should be clear and concise and easy to follow. These would be built to support an environment focused on safety, quality, and repeatability. People, in general, are creatures of habit which makes us susceptible to following a given standard. Therefore, when the standard is written based on these types of parameters, it starts to build a sustainable system. The standards are our baseline to build upon, without having created a standard you have no way to understand if a change is an improvement. These standards are foundational building blocks to the system of a company working to become a premier performer. Additionally, these standards lead to the company's ability to interchange its people in and out of different roles to continue the development of people also allowing new sets of eyes to improve the standard, creating a continuous improvement cycle and culture.

Leader Standard Work

Leader standard work is another area we tend to leave out. If you go on vacation and do not have this for others, how do we know they will follow up on all the items that need attention? When you get back from vacation, if you fail to lead with standards, everything is a mess. This is an area we should lead by example. Leader standard work will also help you ensure that key items are done when expected to be done. I started to understand the need for leader standard work when I became a leader. I didn't have any standard work from the previous leader and had not developed any for myself at the time. I was trying to manage my day from my Outlook calendar. It managed my meetings but was not managing my reports or my follow actions to projects or my people. It also was not helping me to self-develop. After several failures of meeting expectations, I realized how important the leader standard work could be. I was able to add the following activities and reports to my standard work and when I missed to follow up or report and I was able to understand what caused the miss. It could have been a crazy day but was no longer because I just forgot. I was also able to add self-development items to my daily standard work. An example would be to smile more often or to stop using a word that was a trigger to others.

I gave you a little bit of my story. Now, there are many benefits to having leader standard work. It gives your day a focused plan, by having key items that need to be done during the day. It gives you the checklist to drive the repeatable audits, reviews, and coaching that needs to be done weekly. It gives a foundation for the next leader that steps in your role to start with. It helps you drive a strategy when the daily firefighting is happening. The leader standard work should be a repeatable process. The higher we go in an organization, you will find that you will have fewer repeatable actions. This does not mean you will have none at the CEO level. Leader standard work drives discipline in your daily approach to leadership. It is visual to the user to understand if they are addressing the key items of their daily work.

At this time, it would be a good idea for you to write down what you see that is repeatable in your day, week, and month. Once you do this you will have a start, as you go through your month and review the list daily, you will find other items that are repeatable that you can add (refer to Figure 4.6). You can add the behavioral changes in your standard work, to be reminders of something you are wanting to develop. This will help you drive repeatable and predictable leadership for all to follow. You may find that you can't repeat the items you have listed hourly and see that some just need to be done by a certain time. That is okay because the purpose of the leader standard work is to give you a structured list of repeatable items that you need to do and by when.

One thing to remember is that standards and standardized work drive a repeatable process. It also is a baseline we can return to and even improve from. It should control defects if the standards and standardized work are followed. With this thinking, we are protecting our customers. By making this a key point in our leadership to follow our standards shows that we respect our people by driving success in following what we know as

Leader Standard Work

Week: ______	Where there is no standard, there is no improvement	X Incomplete ☑ Complete

Mon	Tue	Wed	Thu	Fri	Weekly Follow-up		Actions, Observations and Questions
					Review Change Agent Std Work		
					MDI Walks 6:45		
					Update Plant board		
					OPEX Team Meeting 9:00		
					Complete Open Concur actions		
					Gemba Walk 7:30		
					Staff Meeting 12:00		
					Corp OEPX Meeting 3:00		

Wk-1	Wk-2	Wk-3	Wk-4	Wk-5	Monthly Follow-up		Actions
4th					KPI Survey		
5th					Send out reminder for Financials and Quaility		
6th					Box 8 Update / kpi		
	7th				Download CEO metrics		
	7th				Update OEDP		
	7th				Update STD Work % in Sharepoint		
	8th				Box 8 Call		
					5s Audit for Campbellsville		
		12th			MDI Walkers calender		
					Steering Committee MTG		
					Verify MDI Walks & Plant Board KPIs		
			23rd		Update OpEx Headcount		
			25th		PMP Review		

Q1	Q2	Q3	Q4		Quarterly Follow-up		Actions, Observations and Questions
					VS 1 Refresh		
					VS 2 Refresh		

M	T	W	TH	FR	S	S	Daily Standard Work		Actions, Observations and Questions
☐	☐	☐	☐	☐	☐	☐	Review tomorrow's meetings / adjust plan		
☐	☐	☐	☐	☐	☐	☐	Clear email inbox / Tasks		
☐	☐	☐	☐	☐	☐	☐	Practice empathetic listening - echo back / nodding / tell them you apprec.		
☐☐	☐☐	☐☐	☐☐	☐☐	☐☐	☐☐	pr / read		
☐	☐	☐	☐	☐	☐	☐	Tell each member of my family I love them		
☐	☐	☐	☐	☐	☐	☐	Genuine praise / recogn. of at least one indvl.		

Notes	

---------------------------- EOW Self Assessment/Comments ----------------------------	
	Say it once, Be concise, KISS - speak at the rate your audience can absorb

John Doe

Figure 4.6 Leader Standard Work.

best. By auditing the standards and standardized work, we drive the go and see. With all things Lean, we can and should continuously improve our standardized work and standard processes. In this thinking, we are following the four Lean philosophies.

Chapter 5

Communication

Why do we fail at communication? Do we not understand that people need to know if they are winning or losing? We don't make our vision clear. We don't give clear directions. If our people don't understand our vision and the direction that we want them to go, then how can we get their buy-in? Lack of understanding can drive people to leave your group or even the company.

Hour × Hour

How do we know if we are winning or losing each day or even each hour? Hour-by-hour or end-of-line boards is a major change in some people's thinking. Some places have all this information in reports for our leaders to see. How important do you think it would be for our people to know their performance on an hour-by-hour basis? I have experienced supervisors not knowing this information every hour of the day. If a supervisor does not see they are losing until the end of the day, it is too late for them to really do anything about it. Operators need to know the goal of the day and how they are performing to the plan. Everyone wants to win and if you are not keeping score then they are just kids on the soccer field kicking the ball around waiting to get their participation award. All the while the customer is saying, why am I doing business with this company.

Hour-by-hour boards should be clear in what the plan is and how they are meeting the plan (refer to Figure 5.1). Red works great when we are not meeting our plan and green is clear that we are meeting the plan. When we

DOI: 10.4324/9781003194781-5

Time	Target	Actuals	Variance	Cumulative Total	Current Part # / Comments / Problems	Owner
7 – 8	23	23	0	23 23		RC
8 – 9	25	23	2	23 46	*Press down for 5 minutes*	RC
9 – 10	23	23	0	23 69		RC
10 - 11	25					
11 – 12	12					
12 - 1	25					
1 – 2	23					
2 – 3	25					

Figure 5.1 Hour × Hour Board.

use colors like red and green, it makes it very visual at a glance if we are doing well or if we have an issue that may need leadership involvement. These boards should be an information hub for the operators and leaders. They can show our daily problems, which ones we are working on, and where people should be in the process. The standard work for the processes should be near the process, or at the least on this board. We have always set our hourly rate based on the bottleneck process or process that is the constraint of the line. We do this because that process is the process that sets the pace. You can only go as fast as your slowest process. Now, if we have done our work upfront that process will meet the takt or pace, the customer wants the product based on our available time to work. You can show the quality issues that are coming out of the processes. This will be listed on the board in a comment section. We should also list short-term countermeasures. The reason for listing this data is to make it clear to all – what is happening and how we are addressing our issues. There are many things we can put on these boards, my suggestion is to keep it simple starting out. People get overwhelmed with paperwork and not all things need to be measured starting out. It is very important that the supervisor reviews this board every hour on the hour. This gives them the ability to respond faster to issues and course-correct before it's too late in the day. Other leaders at the site need to review them at least twice per day. When leadership is at the boards, it shows it is important to the operators, especially if actions are being driven from these meetings. It should not be about who made the mistake, it should be about how can we help correct the issue and make the process better. Learning should be a focus along with the delighters. When I talk about delighters, I mean what is the extra we can do to make a process better.

Andon

Now that we have covered some hour-by-hour boards, let's look into how we can help the operator get help at the time an issue is discovered. We need to put a system in place so the operator can get help before the hour is up. This system is called Andon. An Andon refers to any visual display that shows status information on the plant floor. Its origin is in the Japanese word for "paper lantern". It is the connection from the team member to the team leader or supervisor. The underlying principle and purpose of Andon are to prevent downtime by calling for help and visually show the status of the line or process. Andon's can be very high-tech to very simple. In the automotive world, you will find the high-tech side with sounds, boards, or monitors hanging over the lines, with data collection going on in the background. I have seen simple solo cups being used. You need to decide where you want to start, I would not let cost get in the way, go simple first. This empowers your team members to be part of the problem-solving and quick response.

In this example, you can see that there is a light for each process on the production line (refer to Figure 5.2). If there is an issue at a process, the team member can activate the Andon and it will turn the light on, with an audible sound. This will alert the team leader to respond to this section of the product line. This is important to understand because it allows for a quick response to abnormal conditions that happen during the day. There are lights for the faults the line may have that would cause the line to stop. Some Andon's would have the current count of units made. You can show downtime and line performance metrics. We don't want to get lost in too

Figure 5.2 Andon Display.

much information. We need to remember that the main purpose of the Andon is to communicate for quick response. By doing so, we will reduce downtime from looking for help. I have seen production stop because a team member had to use the restroom. They would just walk off the process and everyone would wait for them to return. When they could have called with an Andon for a team leader to relieve them, without the production stops. The same for any abnormal condition, like equipment performance failures. By using the Andon, nobody leaves the process and a call for help goes out. We also need to think about the team members, when they leave a process and are not in their regular standardized work, this is when defects can happen and someone can be injured.

One of the mistakes people make about the Andon is they want to collect data, so they can problem-solve, they make that their focus. That can be a benefit, but don't make data collection the main focus. Instead, make the process of using the Andon system a focus. Data collection is great and very important to problem-solving, but we want to reduce waste now. Let's cover the process for how the Andon should be used by our front-line workers. When a team member discovers a problem, they should activate the Andon. This will send the signal for assistance. The team leader should respond and evaluate the situation. We want this response to be fast, so understand your span of control for the team leaders. After the team leader assumes ownership of the problem, they can release the Andon or keep the Andon on. The team leader will then correct the problem or call for help. We need to remember that the team leaders are a conduit to support groups, they are the ones that call for help as needed. They are also the person that may make a call to continue to keep a process from moving forward until the escalation process moves to the next level leader. If we are using the Andon properly, we should see less downtime because of escalation process is moving faster. How many times have we watched a process just wait for someone to be found to make a decision? How many times have we seen a process stop because someone went to the bathroom? If a team leader was called to the process, they could choose to replace the team member while they are away. The Andon gives better control for leaders at all levels to respond and make a decision on what is best at that time.

As for collecting data, we can tie the Andon into equipment for detailed data on the performance of each machine. You can even have a maintenance call (Andon for maintenance) added to the equipment for faster response. There are systems we have used that give live performance data of the line or cell. We have used Andon systems that we can see remotely or even off-site. You can store this data for problem-solving. The sky can be the limit for data collection, just remember that data collection is not the main purpose for Andon, it should first be a way to signal for help.

Opportunities to improve and respect for people are critical for the success of an Andon system. I remember when I first started at Toyota, we were in a big growth mode. Adding a second assembly plant and hiring 100 team members per week. We were also making about 525 cars per shift. This put us all in a very stressful situation. The production line I was on had 20 processes and 4 team leaders. The team leaders were on a process 50% of the time because of team members being off work or on restricted duty. So, on any given day, we would have only 2 team leaders to support 20 processes, 10 for each of the team leaders. Because of all the team members being new, we were having about 600 Andon requests per night. That was about 300 Andon responses per team leader. Remember we are only making about 525 cars per night. So, the team leader's workload was heavy. One of the key things I remember was how the team leader would respond to an Andon. They would respond with respect, "How can I help you"? There were other questions that followed like, "Why did you cross thread the bolt"? This one threw me off at first. Why would my team leader ask this question? The answer was he wanted to engage me in problem-solving, get me to think about how I can help correct the defect. After a month or so of this type of problem-solving with all the team members, we reduced our Andon pulls from 600 to about 100 per night. I'll never forget this type of respect and problem-solving and used it as I advanced. People want to win and be part of the winning process. By treating them with respect and engaging them to drive continuous improvement, we all win. The Andon and the hour-by-hour boards go together well. I will add that the best Andon's are visual for status at a glance and are auditable for all to hear until the response is made. This promotes the go and see behavior.

Alignment/Buy-in

Communication is a key part of Buy-in. We should do our best to help everyone understand the key points and the reasons why we are wanting others to do something. Communication plans are a great place to start. This is a simple plan on who, when, and how you plan to communicate. By doing so, it makes you think about all the people that need to be involved. When deploying a major change in the organization or even a minor change on the floor, we need to understand everyone that needs to know how we are planning on communicating, and when. Communication is always a low score on any engagement measurement. We cannot forget to explain the why. When I was an operator, I would be upset when leaders kept me in the dark about a change. I always felt it was best to share news even if it is bad. I wanted to be part of the solution, not just part of the problem. Don't forget the A3 as a way to communicate. With the A3, we state the reason for action, our current situation, and where we want to be. It gives detailed information on how you did your analysis. The solution to moving forward and a detailed plan on how to make it happen. It will show our progress and what we have learned along the journey. If your company is measuring engagement or if you want your people to be on a team, then treat them like they are part of a team and engage them, you may be surprised at the results.

Communicating through training is a great way to get an understanding of why we are doing what we are doing. I used to issue weekly toolbox talks. They were simple with a short message on a subject. An example would be 5s, I would explain all the parts of 5s and may even give a quote from a famous person. Try not to use Japanese examples for quotes all the time. There are great leaders from all over the world for this. The toolbox talks would be given out at one of the morning meetings at the operation. This would be done with company information like days off, policy changes, and so on. The message here is don't miss opportunities to train. Plant communication monitors are great also. Place a monitor in the break area and run a PowerPoint slide in auto showing the same info being communicated at the process. You can also show pictures of your people at picnics, retirements, new hires, and so on.

Mission statements and values are very important to have in front of all of your people and yourself. We all need to be reminded of why and how we are to do work. People are very funny about how they can lose focus on the why and how. Is your business customer focus? If so, do we all understand how we are going to serve our customers? Is it alright to bypass a safety device to serve the customer? You would say of course it is not alright to do that. Well, I can tell you people forget this all the time or we would not see people getting hurt by doing this.

Communication is an action that most of us generally feel we are successful at doing. All the while, we often miscommunicate or simply fail to communicate a given message. This takes place at all levels of leadership in most companies, from an enterprise-level announcement to on-the-floor communication about an immediate issue with the product. Why is communication so challenging or perceived to be a failure point by many people inside a given company? Do we overcomplicate the message? When we are tasked with deploying a communication, do we clearly understand the intent of the message, not just the message itself? Understanding the intent allows the tailoring of the message to become apparent or clear. It's been my experience that tailoring the message needs to start with WIIFM (What's In It For Me). Each audience is immediately translating the message into "how does this fit into 'MY' world, why should I care about this communication, is this good or bad for ME". One might say that knowing the audience and its WIIFM is, in fact, one of the most critical pieces of the puzzle of communication.

Communication is the act or process of exchanging information between individuals by way of verbal, symbols, signs, or behaviors. The keyword in the definition in my mind is "exchange", meaning this is a two-way street. Giving information and receiving information and vice-a-versa. Often in my experience, I have been coached to listen more and talk less, "you have two ears and one mouth for a reason". Now, I find myself communicating that very comment to the ones I now mentor. In terms of running a business, we must work to identify best practices for effective communication using multiple media and human senses such as sight, sound, and potentially feel. Let's dive deeper into some methods that I have seen and used to help make a more effective communication path.

Leadership to the Next Level

A plan for development is a great way to connect to your people. Leaders should sit down with each of their team members one-on-one to understand what their goals are in their career. Having this discussion can clearly lay out a path to make this goal happen. When I first started in a leadership role, my leader developed a path for me. This was a good thing because I had no clue what I needed to be developed at this time. He exposed me to many roles and situations that gave me the experience I never knew I needed. I also remember a story of a leader that thought they were happy in their current role and had no desire for advancement. They met with a great leader that would not accept this for an answer. After some questions about what they love to do versus what they dislike to do, they discovered a new path in their development that would lead them to a more meaningful career. It is important to understand what makes your people happy and where they want to go in life. If we understand these goals, we can clear a path for our people to be successful.

Expectations should be clear to all our people. If we don't communicate what we expect, we are sending a message of no vision or direction. Expectations for meeting metrics are not a clear message. There needs to be a strategy to move forward. If the ship is sinking and we must move fast, it needs to be understood by all. We need to also state the future after we fix the leaks and how we will move forward. This gives hope that there will be stability. Which metrics are most important? Why would this need to be known? Everyone has their view of the world and the team needs to have common goals. Your team can and should work to improve all metrics. With this being said, some metrics may need more focus than others. I had a leader tell me once if we can keep a certain metric green, it gives us permission to do the other things we want to do, to move forward with our people.

Regular open discussions are important to develop and maintain alignment with your team. The regular pace of open discussion can keep your team on the same page, without the micromanagement that most people dislike. There also needs to be one-on-one time with your people. During this time, you may find that an individual will open up more. It is also a better time to talk about personality improvements.

Understanding your team's personality traits will help you in your alignment and teamwork. Using a personal test will give a better understanding of who may need more attention and how to approach their needs. I remember taking a personality test and had my eyes opened to why my team was struggling with me and each other. Some people can be very dominating, wanting to move fast, while others struggled with this because they were very analytic and needed more information to move forward. Once you and your team start to understand each other, they can adjust how they interact with each other. I have seen some companies where they wore their personality trait on their company badge. This gave each team member a quick visual indicator of that person's personality traits.

Being an awesome listener is a great step in being a great leader. Communication should be a two-way street, not just a bullhorn telling your people what to do. We should listen with the intent to understand, not just respond. Being quiet, and letting your people speak can be eye-opening. It is always best that we speak with facts and data, the problem is some of your people may be struggling personally and emotion takes over. When this happens, we need to listen and understand the personal struggle. Sometimes that is all that is needed, a time to vent. Other times they are looking for direction in the personal side of the problem. We need to know the difference; it is okay to ask if they want to just vent or is there more?

Body language is another form of communication that we as leaders need to be looking for and understand. You may find that a person is saying one thing and the body language is saying another. Don't ignore this, dig in and ask thoughtful questions to drive further dialog with the individual. We may find out that your message is not being received well and needs more detail or adjustment. Our body language is very important when we communicate. If we are rolling our eyes while explaining a change coming from other leaders can send a bad message.

With communication, we are connecting to our customers, which are our people or co-workers. When we do this well, we are engaging them and this shows respect by making them part of the solution. If we are making our communication a two-way street, it can drive improvement. Always go and see to fully understand our customer's perspectives.

Chapter 6

Leadership

Why does Lean fail when it comes to leadership? Is it because leaders don't understand the Lean philosophies? Is it because they don't have a vision or aren't able to articulate the vision? If leadership has a vision, are they able to remove any roadblock that restricts conveying the vision and gain alignment with all their people? Maybe the leaders learned some bad behaviors and they reflect behaviors more of a manager's than a leader's. Changing with the times can be challenging for people in general and more so for leaders making key decisions for the future of the company, defaulting back to what we know is the typical go-to behavior. So, do they know they are leading by example, even if it is bad? How do our leaders respond in good times and bad times is a question that comes to mind?

Team Leaders

Let's start with the team leader. This is typically an hourly person that is supporting approximately five to ten team members. This position was designed to be an off-line position like a supervisor, in fact, if you are training and developing them correctly, they are your next supervisors or group leaders. The key training areas you need for the team leader role are train within industry (TWI), standardized work, problem-solving, waste elimination, communication, and servant leadership.

TWI is a foundational part of their role. TWI is a method for training people on a new process. This was developed in the early 1940s by the United States War Department. We will cover this method in more detail

DOI: 10.4324/9781003194781-6

in the next chapter. The reason it is important for a team leader to use this method is that we see too many times in today's industry, people are not being trained correctly on their processes only to find out later that performance is suffering. You will hear me say this again, a company pays for training in two ways, on the front end or the back end. Meaning, we give good training to our people at the beginning or we re-train them and do re-work after the process. So, without proper training, we also find it harder to problem-solve. This is where large amounts of variation in the process show up when people are not at all following standardized work. TWI is instrumental and a foundational way to train people to quickly learn to do a job correctly. "Same Job, Same Way, Every time".

Standardized work is critical to have a safe, repeatable process and the team leader should be skilled in making necessary changes to the standardized work as needed. They are the closest leader to the process and actively work with any issues that may come from our daily work. It only makes sense, if they are training to the standardized work and using the documents to problem-solve issues in the process, they should have the ability to change and develop the new standardized work. We also want the team leader to be able to identify waste in a process so they can remove it. This process of identifying waste for the team leader should be a daily activity that would support the creation of very efficient processes. Additionally, as our team members observe the team leaders, they too will develop without realizing it as they will be immersed in a culture of continuous improvement.

Problem-solving needs to be a strong focus for our team leaders. Most opportunities that a team leader will face will require a simple fix or correction. Being that close to the processes, we need them to understand the difference between a correction or a simple fix and real problem-solving. Otherwise, they may only be firefighting and never find the root cause of the problems they face day-to-day. By developing our team leaders in A3 problem-solving, we are developing future leaders. They will learn critical thinking and communication skills that will support a better conversation and allow them to generate a stronger standardized work document. These critical thinking and A3 problem-solving skills must be a part of every level of leadership. Additionally, it is key for your senior leadership to be able to coach the lower levers of leadership in these skills.

Waste elimination will be a major part of the team leader's role, once we have stable processes in place. The team leader will be trained to perform a similar role to an industrial engineer. Throughout a team leader's day, they are in close contact with several or all of their areas processes, they need to understand how to see waste and be empowered to remove it. This should be directly connected to their yearly review. Once their waste elimination skills are developed, they will even be able to help identify waste in other processes that interact with their processes. For example, I had a team leader explain to me about some waste they saw every day that they felt they could not change because it was not their process. The team leader explained how an office person would bring the daily schedule out to the production lines. During this process that person would stop and talk, interrupting the process flow. So, after some discussion, I asked what do you think could be different, what is a better way for them to get the information they needed without spending a lot of money. After a day or so, the team leader came back with the suggestion of putting a printer lineside. This way they could get the schedules at the exact time it was needed and manage it themselves. This would free up time for the scheduler and reduce the line interruption from socializing with people on the lines. This led to another idea we discussed; how could we use a Kanban/visual signal for some lines instead of a schedule. This is just one of many stories that should involve our team leaders.

The team leader will be a quick responder to issues at the production line or area they are supporting. This will make them a conduit to others that will be supporting these processes. So, we will need to be developing their skills in communication. Talk to the facts, data, and standards, not emotion. By making them the conduit of information to others, it allows the team member to stay on the process as well as "in the know". It will also make the team leader the decision-maker for what to do when a problem is encountered.

There are many other tasks that will be put on their plate as their development is cemented and they embrace the culture. Oftentimes leaders mistakenly use team leaders as manpower or vacation fill-ins and classify them a "working lead". You need to think about how they can do all the above if they are on a process all the time? There is nothing wrong with them working on the line from time to time. It is good for all of us to learn and do a process from time to time, but at what percentage? Any given team leader should be able to step in if a team member has to go to the bathroom or be away from the process for any reason. This will keep the process from stopping the line. I have had leadership say we have team leaders, but they are online team leaders aka "the working lead". Our response is that they are not real team leaders and more like higher-paid team members. The reason for this is that when they are online all the time, there is no opportunity for them to do what we really expect a team leader to do. When they are on a process all the time, we want their focus on the process only and following the standardized work, not thinking about all the additional tasks that need to be done before they can leave for the day. If my mind is not on my process, it sets the stage for mistakes, this is not a healthy or productive environment.

The team leader is a key part of the Lean journey. Developing them and empowering them to make change is the desired culture and foundation for continuous improvement. Understanding the philosophies of Lean and how they will shape their role in the company is instrumental for their success. If we follow through on these items, we will have team leaders that will be great candidates for the next level of leadership!

Group Leaders

Supervisors or group leaders would be your next level of leadership. This role would be a salaried person that may have been promoted from a team leader. This person will be the first level leader that is responsible for human resource issues and responsible for the production of our product. This role is a key player in the development of our people by setting the behaviors that we want our workforce to emulate. This role is very important to manage and ensure they have a clear understanding of the company goals. Just like the team leader, the group leader needs training and development in several key areas such as coaching, problem-solving, empowerment of others, human resource policies, daily management boards, communication, and the values of the company.

Group leaders need to understand the coaching model. If they coach their people versus directing their people, you will find that more people will be engaged in finding a better way to do their daily work. It will drive creative thinking to be actioned. It will also take the burden of having to know all the answers away over time. We will go into further details in this chapter on the coaching model.

The group leader will need to expand their skill in problem-solving as well as their strategy in solving their daily, weekly problems. What we are saying is they need to think about the development of their people. Who will be the next team leader? Who will be their replacement? This role like the team leader role would be intimately close to the waste and problems related to daily production but would be working on the problems that can't be solved by the team leader. For example, they may be building a business case to replace a piece of equipment that causes a lot of pain. They would have a bigger picture of the manufacturing process, than the team leader. At every level of leadership that role should have an elevated view of the production system. If the team leader's view is at say 1000 feet, the supervisor/group leader's view would be 5000 feet, and so on up the chain. Seeing and solving problems at the right level.

Empowering their people is tied to the coaching model. We want our people to feel and know they have the power to drive improvement. It's not about giving up control, we still want to control all changes so that we protect our customers. This is about having a workforce that is driving improvements and understand that we want more from them than just making products. This is a way of showing respect to our people. It shows that we value their thinking and their time at work.

All group leaders should fully understand and follow the policies for human resources. This sets a fair workplace for all. A part of human resources is the development of people. Identifying and working with eager team leaders that are wanting to advance in the company will become their replacement in the future. Disciplinary actions and knowing when a policy is not being followed is a part of the group leader's role. We believe in most cases, if we are developing and showing respect to our people, the reasons for disciplinary actions are reduced. Managing vacations, call in's, and many other human resource tasks will always be part of the group leader's role.

Daily management of the boards and people. The group leader will need to work with the team leaders hourly at the daily management boards reviewing the hour-by-hour portion to understand the performance of the line or cell throughout the day. This allows collaboration between the two to make a decision. If we are on target for the day no course correction is needed, but if we are off target some decisions can be made to get back on target. This will be a coaching conversation to drive corrective actions to any issue the area is affected by. The supervisor or group leader should also have a board that sums up their area's performance to discuss with the support teams and management, review and drive actions in relation to issues. Let's make this clear about our communication boards, they are not just about reading the news. The boards are a tool to help drive support and problem-solving.

We have seen operations when the supervisor/group leader does not manage their people and processes hour-by-hour. We find over time that failures happen or performance declines. Then behaviors trend to the negative from lack of support and leadership. I will never forget, once when I was doing a time study of a team member that was mad. He said I'd like to show you something at the hour-by-hour board. As we looked at the board, he pointed out that they had been running the same part since yesterday and that his rate or pace for the hour was different than the other two shifts. On average, he was hitting his rate of 50 units per hour while the other shifts made their rate, but the rate was half of his. He also noticed that someone posted great job for the other shifts. He said, "that will not be said about me because I missed the higher rate a couple of times today". Think about the behavior we may be developing with this team member. He was expected to make 50 parts per hour and the other shifts were only expected to make 25 parts per hour. Little things like this can make or break us with our people. If the rates change based on a part number, we need to make sure everyone understands the different rates being managed on the hour-by-hour to make the correct rate. If they can't make the rate, we need to help remove the roadblocks so they can have the feeling of winning and the support to make it happen. In this case, not only did we have a team member upset about having to work overtime while we did not hold others to the same standard, but we also lost a shift of production because the two other shifts rate was half of what they can do. The group leader needs to review and work with the team leader hourly at the hour-by-hour boards. Collaborating to drive to the standard and drive improvements to any loss in the standard.

Communication is an area that our group leaders need to be strong. Just like the team leader, they need to communicate with facts and data, articulating the company's vision, goals, and why they are important for all to understand how their work fits into the big picture. Communication is not just telling (reading bullet points), it is also selling (using your personality) to deliver a good message not just "THOU SHALL GO DO". So, it's important that they are trained in communication skills, because if they don't understand and can't explain the direction the company wants to go, then our people may not clearly understand the need for change.

All of Us as Leaders

Earlier in the chapter, we mentioned servant leadership, so let's go a little deeper on this subject. We have seen in our past many leaders that don't understand this idea and are not able to make the connection back to our people and demonstrate the behaviors you would expect to see in return. Leading by example and living the values of the company are good ways to show our people. It is not fixing everything that is brought up. It is not top-down; do as I say leadership. It is not about being soft and letting everyone do what they want. A servant leader leads with others in their mind, they seek opinions (seek to understand), the servant leaders are thinking about how I can develop others, shows compassion, offers encouragement (cheerleader), they are humble, thinks long term, and seeks buy-in from all. A servant leader shows respect to their people by making sure the day is filled with value-added work. They suggest change and encourage the team to drive improvements in their processes.

Experts and coaches are needed to help us along the way. If you are on a Lean journey of transformation, don't you think you would want some of the best to share their knowledge to help your team move faster? You would not start a production line without the correct people or equipment that is needed. The trick here is finding the right someone that fits the mold and brings value to the workplace. Do you need a retired veteran from Toyota or an engineer from one of the big 3? The question you should be asking is what are their roots in Lean, what type of Lean initiatives have you participated in? How does this person relate to people on the floor and are they able to teach and facilitate Lean tools to others? Not all people in the Lean field can do all these things and do them well. The best way I have seen this done is through having a diverse workforce. I am not talking about race or sex. I am talking about someone that can lead, teach, coach, understand the different areas of business, know how to problem solve, know how to get their hands dirty to make it happen. Sometimes this is more than one person to make up a team of different strengths. Some people only know how to tell you their ideas but lack the ability to bring the idea to fruition. A good Lean servant leader will know how to adjust to the environment and the level of understanding of Lean in a company. You need a team that can work together and learns from each other.

One of the biggest killers of a Lean journey is upper management. It is not always their fault, most of the time they don't understand the purpose of the journey and how it is made. They may not be servant leaders. So, it will be very important that they are brought on board this process and that the training is given to all senior leaders. This should not be a flavor of the month; it is to be a journey of transformation. Although sometimes we need to plug the holes in the ship, so we don't sink and go bankrupt. The end game should be transformation and your people need to understand this. This transformation would be about the culture of the people and their role and support to the production system. I have seen in my past when a CEO leaves, the thinking, and direction changes 180 degrees. The problem with this can be that people go back to what they are used to, and the company is in for the worse in the long term. We know companies are not Toyota of today, most are Toyota of 60 years ago. Most companies do not have the clout of a major automotive company. What we all have in common is we have people that can learn and build upon what they have experienced. I have found that the principles of the Toyota Production System and the Toyota Way will work in any environment. While it is not one shoe fits all, it is how you will apply the methodology of the principle in the current environment you are working in. Does it have to be perfect? I would say no. On this journey, a company's leadership team should understand the four Lean philosophies and share that thinking process.

It's key for leadership needs to learn how to coach and understand the benefit of coaching. Coaching is not an easy thing to do for most leaders. Most leaders want to give direction and hold people accountable for following said direction. Coaching is a learned behavior that, as I have seen, pays great dividends. It is a process of seeking to understand what is needed in a given conversation or situation. Then being able to know what can motivate a person or group of people, following up to ensure they have a clear understanding, and encouraging people to move forward. There are skills that a good coach needs to have in their everyday life. They need to have the ability to build trust with people, so they can have stronger relationships to be open with people's development. As a person on a development path, I have to trust that my coach is leading me in the right direction. A good coach needs to be able to ask thoughtful questions to get their people to think on their own to solve a problem. A coach also needs to actively listen, not just listen, but to hear what is being said to them and internalize the conversation before responding. Often our people struggle to communicate the real problem, and by pausing to hearing what is being said, you can drive down to a root cause with some great discussion. A great coach is a person that can get their people to learn and work at a higher level without having to give detailed direction all the time. People start to think on their own knowing the intent of what needs to happen and bring us better solutions than any one leader can do by themselves. You can see in the coaching process the flow of the thinking process (refer to Figure 6.1).

Figure 6.1 Coaching Model.

When looking at this model, it starts with exploring a person's needs. This leads to a conversation to motivate that person to want to win while following up to gain continued alignment and encouragement is key for that person's growth. As a leader follows these three concepts and actions, the arrows you have seen are the by-products of these interactions. This in its-self is a simple coaching model to engrain into your daily habits. As these habits grow, they eventually become beliefs and default behaviors. All of this culminates in building good relationships.

Relationships are a major part of being a leader. Going to where the work is being done and seeking to understand is a big part of this. When you build relationships with people, they are more comfortable in saying how they feel. They also become more open and susceptible to thinking of new ways of doing their process or changes to their area. People need to be free to speak up if they don't agree or understand. A good leader should be able to talk to people and teach them, coach them in ways we can truly learn. We are not always right as a leader, we may not know all the obstacles that may be in the way, understand that, and be humble. When you lose in the relationship area you will lose credibility, when you lose credibility, you lose your people. This is when people will leave your team and leave the company. Who wants to work for someone that you can't trust? If your people are free to be part of the thinking process, the more likely they will be making good choices for the company when you are not present. I have always told my teams; I need people that can think more than I need people that just take direction. I cannot nor do I want to be the answer man for everything, which would be slow and not next-level thinking. Our people need to be able to understand the intent of what we are trying to do and be able to make changes that follow the intent that leads to winning.

People need a leader they can trust. They need a leader that plans for the bad times and looks for opportunities that we can learn from to be better. I remember in 2008 working at Toyota during a bad time for many of us. Toyota had a philosophy in place from the day I started that if times got bad, they would start cuts from the top before they would lay people off. They stuck to this methodology; they had built a strong company that was not debt constraint. Cutting pay from the top first and sending all the contractors and temporary workforce out of the door to save the full-time employees from being hit at this time. All of this too was done to protect their people. Not all companies are like this, but how the leaders react during these times can drive loyalty or it can go the other direction to the loss of trust. I have worked at a company during bad times and the top leadership reacted poorly, with my direct leader making it clear to me how they were trying to protect me and others. I have worked at a company during bad times when the top leadership reacted poorly, with my direct leader making it clear to me how they were trying to protect us. Having open two-way communication during these times is critical to keeping your people. On the flip side, leaders need to recognize and reward when they are doing well. People need to know that they have done well and how they are growing, this will drive them to welcome continued guidance. I have always seen a leader as a parent and the other as a child. It is our role as leaders to be good parents and to develop our children to be great adults. This makes it personal to me how important the role of being a good leader really is.

Leaders should always want to learn and expand their understanding. This will make them stronger and help give their customers better confidence in their ability to lead. The development of their selves is important and the development of their people is critical. An organization of thinkers and learners is a strong organization. People should be taught that problems are just opportunities to get better. Every company will always have problems, it's how we manage those abnormalities that are critical. This reaction to the problem is key. Don't be a victim and just accept them. Develop a plan that protects the customer, while being respectful to people and driving real improvement to processes that will make us stronger for the future. Lots of growth should be happening when we are dealing with our problems and opportunities.

It does not matter what level in a company you are if you are a leader, we are all expected to lead by example and living by the four Lean philosophies. Understanding the value of coaching others as well as being humble, seeking to learn and develop ourselves, and not waiting for someone to develop us. We need to have a vision, a true north statement so all can understand that this is the direction we will all work toward in all that we do. The vision needs to be breakthrough thinking. This means goals should be stretched higher than most believe we can currently do. I always remember a goal not set high is a goal not met. The goal should be attainable based on your team's skill set. Sometimes we need to develop our team to make the higher goals and we need to understand what is needed for development.

Once the vision is set, we need to have a path to get there. Strategy plans for each area of the business to meet the new goals. What do you need to achieve this year and a vision for the next three to five years will help us develop the strategy? This needs to be reviewed weekly, monthly, and yearly to understand how we would need to adjust and correct our course.

The reason for action matters. I remember an event I participated in and the problem came up about a call center being overwhelmed. A few months before they had gone through a headcount reduction because of a goal to reduce people across the business. They felt that there were not enough people to handle the calls, and that was the problem to pursue. I asked the question, "why are we having so many calls"? They had some data to share, and the response was failures were high on a certain part from a poor design. The cost to correct the part issue was higher than just replacing the parts when they fail. So, the team felt they could not correct the design of the part because of the high cost of redesign. The math was correct but was this the right thing to do for our customer? Would this cause a loss of new sales to these customers or other customers that may have heard of this issue by word of mouth? If we only look at cost, we may not be doing the right thing. Sometimes, we need to build a good business case to serve our customers better. We could have added more people to the call center to be quicker at the response. If we had done that, we could claim the extra headcount as part of the savings later. If we try to add that savings upfront for adding headcount at the call center, we would not get credit because it was cost avoidance. With this thinking, what behavior are you driving? What behavior do you want to drive?

What we measure can drive good and bad actions. A key item for leaders is to understand financial accounting for Lean thinking. I was working with a controller once and he was pushing absorption metrics in the plant. It was the first time I had heard of this metric and wanted to learn more about it. To do well in this metric, we had to build as much as we could and if the cost standards were better for a certain product, we should make more of that to help drive the number higher. The problem was, we had no orders for this part with the higher cost standard and we were overproducing units. It made no sense to me and his response was, “we get credit for the work and we keep our people working”. The real issue, credit for work that receives no cash until a customer buys it is a real problem. To add to our pain, the problem came back at us in another form months later. I saw the same units come back into the plant to be reworked because of rust and UV damage. More cost to this product from movement, rework labor, material cost, and the risk of having a poor-quality product because of not going through a standard process with controls. This is why overproduction is the worst waste of all. So, when it comes to accounting for Lean metrics, they should be customer-focused, not asset or labor credit-focused. It may be good to watch these numbers but they should not be key metrics to drive the business. Overtime, overhead cost, freight cost, and other financial metrics are more important than absorption cost. First time right, on-time delivery, schedule attainment, parts per man-hour, and inventory turns are a few other metrics that are more customer-focused. Always remember that what we measure will drive good or bad behaviors in our business.

The four Lean philosophies are key to being a great leader. If we apply them to our day-to-day operations, it will make us predictable to our people in what we expect. Our people will understand better how to apply the intent of an idea that will be more aligned with your thinking.

Chapter 7

The Why

Why do we fail at doing it right the first time? Why do we fail at understanding how we can adjust based on our needs and still follow the intent? Why do we find people going off into left field and never get close to what was expected? It is most important for people to understand "the why" they are doing something. Think about a soldier walking onto a battlefield and not understanding the real reason they should be there and why they should do battle. If they have a greater purpose and meaning to win than to just survive, the more they might put into the battle. I know this sounds extreme and most tasks are not seen as a battle for your life at work. I would say if a person knew the deeper meaning for why they need to succeed and why they are doing a task, the more they can put into a task. Also, if the person knew more about the why, the more likely they could make the right decision when an issue comes up. They may be able to think like you as a leader or even make a better decision because they are closer to the problem. It's my opinion that when people have the "Why" they are now more accountable for their actions! They can no longer say they don't or didn't understand! I was recently asked by a VP what would the "Why" look like in our work instructions? I replied with it can be very simple as long as it provides a clear understanding, for example, we as a group had just been discussing a rough edge on a ceiling panel. I explained that by adding to the work instructions the "What" I'm going to do, which is inspecting the edge of the ceiling panel before installation. Now, the "How" I'm going to do, run finger along the edge to confirm it is smooth with no imperfections. And, lastly the "Why" I am doing this activity because it is a customer-facing edge and very visible to all who pass it. So, many

DOI: 10.4324/9781003194781-7

people may say that this is "common sense" and it's obvious that the part should not be used if it has a rough edge, but common sense translates to subjectivity and allows room for misjudgment. Now, as an operator, I know "Why" I should or should not use the part that was provided to me. It's a customer-facing part and if it has a rough edge, I will not use it.

Another reason to explain the "Why" is when we are problem-solving. Problems have a why and it is up to us to drive down to identify the root cause of each problem and, therefore, the why the problem occurred. Does it make sense to try and solve a problem without understanding the Why? I cut my teeth when I was working at Toyota using the term 5 Why's. 5 Why's was designed to drive us to think deeper. It is usually easy to think through two or three why's as you work through a problem. Fourth and Fifth why's were often more challenging to think through, but also the key answers to get to the root cause.

For example: Problem: Seal Defect
Why: Smeared seal
Why: Using the wrong technique
Why: Standardized worksheet (SWS) not accurate
Countermeasure: Update SWS

This, in my experience, has been a typical response when people are learning to use the 5 Why's. While this might seem reasonable to many people, we went wrong on the second why. What about the technique that caused the smear? Identifying the actual actions that the operator used to create the smear would be the correct path for our why's.

For example: Problem: Seal Defect
Why: Smeared seal
Why: Dragged elbow across the seal bead
Why: Arm was in the wrong position
Why: Using the wrong hand to hold the tool

Now, while I still did not go 5 Why's deep, one can see the difference in the deeper thinking. Diving into the details to drive to the true root cause is the difference between solving a problem and firefighting.

I will agree that not all problems require 5 Whys, but I will never agree that a "Why Chain" will only be one why. For example: when a socket on a pneumatic impact gun cracks and breaks, you just make the decision to change the socket. Case closed, no need to contact a metallurgist to try and identify why the steel socket broke after being used for 10,000 shots. Now, if the new socket breaks within a week or so, one might start to ask why. Is the socket rated for the gun? Has the air pressure changed? The list can go on and on which could lead to more than 5 Whys. And for this reason, at my current company, we use the term "Why Chain" versus "5 Why's". The five sets not only a target but also a limit. While we need numbers in most instances, we still need to be cautious not to create boundaries or limits that will stifle our thinking!

Validation of Understanding the Why

One item that is often missed is validating that our people understand the task at hand or simply the expectation. When do you know someone knows? When you ask the simple question, "do you understand" and the person states "Yes sir I understand" are they able to set you at ease, or do we often feel uncomfortable that clarity was met? In many cases, especially when a person is new, they want to show their boss they can do their job and do it well, so asking, clarifying questions is OUT of the question. I, myself have behaved this way before, feeling as if I must show how capable I am and I can figure things out for myself. But in reality, had I just asked a few more questions, I could have saved myself and the company from the time that I wasted trying to work things myself. My point is that it is critical to ask probing questions while reading the body language of the person. The questions would be the piece that requires the person to repeat back what they understand to be true, and the body language would be the piece that displays the person's level of comfort in their understanding.

Once our people understand the why they are doing something, the more likely they will be able to adjust to the situations that come up and follow the intent of the task or assignment. I have seen so many times when people don't fully understand the why they are doing something, and the results are far from what was planned. When they understand, it may even exceed what was planned. Transformation can take place with people when they can adjust to knowing the intent. True change in thinking and learning begins. Understanding the intent and expectation is critical in getting the best results. There are many times we have been in the middle of a project and things don't go as planned and we have to make adjustments. Would it not be better for all to understand the intent and have the ability to make changes that are aligned with the intent? Think about how much faster and how much better the end results can be if all knew the intent (the why) and had the authority to make changes in this spirit on the fly.

I remember a time that our leadership started a program that all sites would have to follow. The problem we found was that our businesses were very different, and the standard rollout did not fit as a copy-paste. For this concept to work, we would need to adapt to our type of business. So, understanding the why and the intent was critical for success. The first thing we did was understand this and come up with our solution. Once we had this, we went to our leadership to gain alignment with them. I have always coached to understand the intent and develop a solution before we go and complain that it will not work. Don't just check the box, develop a better way to win.

Key Points for Why

Standardized work is a place that many forget to include the why. Training within industry (TWI) teaches us to cover the what, the how, and the why. TWI was developed during World War II, by the United States Department of War. The purpose or why they developed this method was to train workers that were making military products. The United States had a constraint of trained and skilled workers entering the workforce to replace skilled workers that went to war, this training provided a fix for this problem. Later, Toyota adapted this method of training and uses it to this day. With these elements being understood, an operator can be a greater benefit to problem-solving when there is an issue. The "What" is what we are to do, a major step. The how is the method (key point) of what we are to do, safety or quality clarification may be called out here. The why is this the understanding of why the step is important (the reason why). The why is where we are driving the expertise into the trainee's mind. I have seen many operators not fully understand why they are doing a task. One example was an operator paint marking bolts before they were to be installed. I asked why they were paint marking the bolts and the response stung. The operator explained they were not sure and thought it was a Lean thing and help the process move faster. Keep in mind the operator was marking the bolts in bulk, not one-piece flow. I explained that it may be because it shows they have been torqued. After some discussion, the operator understood what they were doing did not make sense and changed to marking them after they were torqued. The "do as I say" thinking does not work and transformation does not take place with "do as I say". It is critical to have the What, How, and Why in any standardized work. This would be the baseline for the best we would know how to do a process.

Below is the TWI method we have used during our time at Toyota and after we have gone out into the world in other industries (refer to Figure 7.1).

The other way to look at training a new person is getting all the detailed knowledge that the expert or seasoned person had while doing the job. We often put new people on processes all the time with very little knowledge. "This is how you put this together, now do it fast". How is this protecting our customers? How is this protecting our new person from getting hurt or creating a defect? If we give the person the "What, How, and Why" with all details explained in the key points of each step and the reasons why we do it this way, we could build up an expert faster by transferring the knowledge in a formal way. This is critical when we are often hiring people

How to Instruct

Prepare T/M

* Put T/M at ease
* State the job
* Find out what the T/M already knows about it
* Get T/M interested in learning the job
* Place T/M in correct position

Present operation

1. Tell, show and illustrate each Major Step one at a time
2. Tell, show and illustrate each Major step, stressing Key Points/Knacks
3. Tell, show and illustrate each Major steps, stressing Reasons for each of the key points
   *Instruct clearly, completely, and patiently
   * Present no more than the T/M can master (Do you have any questions?)

Try Out Performance

4. Have T/M do the job; correct errors
5. Have T/M explain Major Steps as the job is done again
6. Have T/M explain Key Points/Knacks for each major step as the job is done again
7. Have T/M explain Reasons for each key point as the is done again

Follow Up

* Put T/M on own
* Designate to whom to go for help and where
* Check frequently
* Encourage questions
* Give any necessary extra coaching and taper off the follow up

Figure 7.1 TWI 7 Step Training Method.

with a low skill set or a person who may have never worked in this industry before. Additionally, even experienced workers need to know the key points and reasons why they doing a given task. This helps them know when the process is not operating the way it was intended and needs to be addressed. It protects our customers.

Buy-In

On a higher level such as deploying a major program or product line, it is very important for all to understand the why. As I explained earlier, it helps people adjust to the problems that come up in their environment. I have seen the leadership of several companies trying to make every location and function use the same tools. An example would be process boards or end-of-line boards. These tools don't always work, since a machine cell, assembly line, functional testing, shipping dock, and other areas will have differences that would drive different metrics. When people understand the why or intent of a process, they can adjust to meet a meaningful process for their operation. Using a Strategy A3 can help us drive this understanding. We want to align our people to understand the true north of our goals and direction. By doing so, they will more likely to align and adapt to their areas to move in that direction. We don't want people to just follow orders, we want them to understand intent so if things don't work out, they can adjust to the intent and get the results we want. This does not mean free range to do what you want. We must be clear about the guidelines and what we want to see. Standards are important when it comes to many areas implementing a process that should be the same. Think about it. If you are a leader going from site to site and you are reviewing the new communication boards (refer to Figure 7.2). Every site you go to has a different format with the same or mostly the same info. It may take you a bit of time to adjust to each new board and you may miss an opportunity to help coach the team to a better level. This is a form of waste (time). Standards bring us to a better state for faster understanding.

Understanding the why we are doing a process and the guidelines or standards that need to be met will lead us to a more successful end. The customer will be happier with the end product. Our people will be happier with being included in the importance of how and why they complete a process. Leadership will be happier that their people were more successful in doing it right the first time and maybe even bring the process to a higher

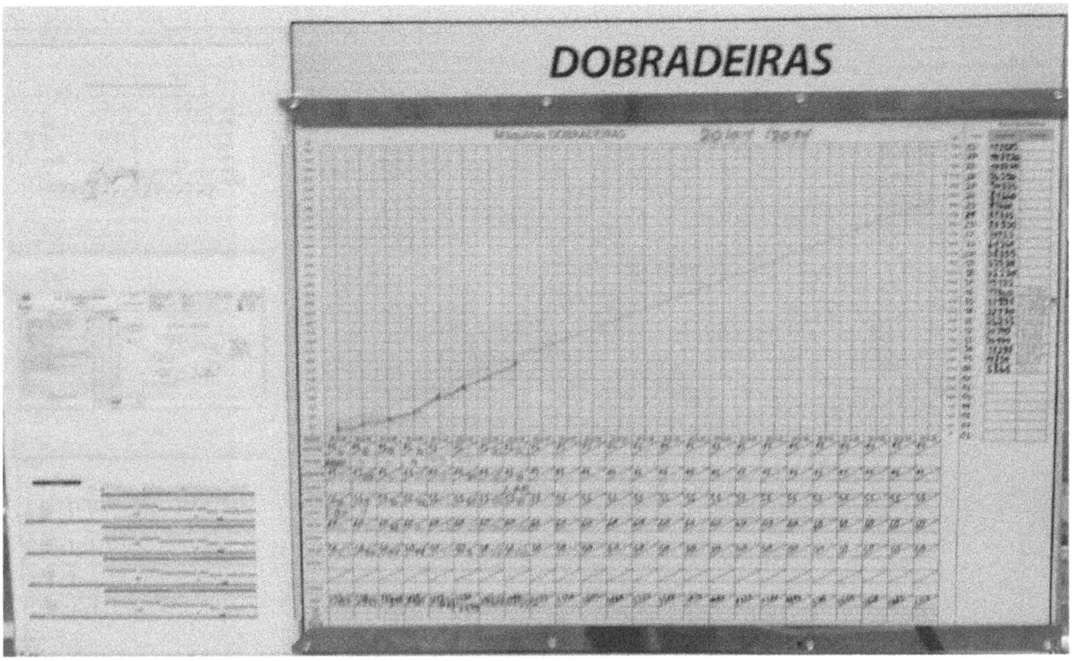

Figure 7.2 Communication Board.

level. In the end, it is about respecting our people by giving them the understanding of our intent and protecting our customers from any of our misunderstandings. With the intent, we are allowing our people to improve the process for their area if needed. Going and seeing the process will drive the four philosophies of Lean.

Chapter 8

Reflection

Why do we fail at improving a process the next time? Why do we fail at learning from our mistakes? Why do we fail at being grounded and bringing ourselves to a higher level? Reflection may be a key factor for the gap in our thinking.

Reflection is a key part of the way most at Toyota think. Reflection is not about who do we blame, it is more about what did we learn. How can we take the learning from our failures and improve? How can we take our successes and build upon them so we can be even better the next time? How could we complete this task faster the next time? What work could we move outside of the event? What could be pre-work and what could be post-work? Are we really winning? Are we measuring the right key performance indicators (KPIs)? How did we work with our people? How did our people do during an event? Did we stretch our thinking? Did we learn from what we did?

After every event or project, you should be asking the above questions to improve your thinking and the outcome of the next project. You should document the learnings in some kind of standard work. If you and your team are doing this, you will build upon each event and be stronger and faster each time. If you put the improvements into standard work, you will more likely not forget it the next time and new people on the team will have the information in from your learning, so they will not make the mistakes from the past. When we started at Toyota, we would see a model change on the production floor take several days to complete. That was several days without making the product. We were setting up our processes for the next model, moving out old material, bringing in new material,

DOI: 10.4324/9781003194781-8

building racks to hold the material, and so on. Before we left Toyota, they had got this process down to 30 minutes between models. This took years of reflection and improvements. Over those years, we had many leaders and team members come and go from the team developing this process. So, documentation was critical to the success of future model changes.

Reflection after problem-solving can be helpful to improve this process for the next time. Many times that you may have forgotten to do something or had someone upset over a change that you made. By reflecting and documenting your learnings, as said in the above paragraph, you can keep it from happening again. There is a lot of learning we can get out of problem-solving. We may discover that our data tracking is poor which leads us down to the wrong path. We may find out that we need certain people to be involved in this process because of their skillset. We may see that our project management skills need to be improved. Without reflection and a plan to improve, we will see much of the same results. Even if they are good results, we should always be looking for better results. How can we complete an A3 for problem-solving without losing the quality of solving a problem with data?

Reflection needs to happen after coaching a person or when a person is being coached. I will never forget being coached about not meeting the standard that a leader had expected. I had made great improvements with the help of my team on a production line. We reduced the overtime on the line by three hours a day, increased the output to meet the customer demand, and reduced the number of people needed to make the product at the same time. The president had a vision of a nice-looking area with floors marked and painted surfaces. The production area was also still in a warehouse area. It was still rough at first sight, but we were still a work in process. The data showed great improvement and the morale of the people had greatly improved. I was told I did not know what good looked like in front of those people and received more comments on how we fell short. I will say it got a little heated when I pushed back on how we were winning and still working on progress. After this, I had to do some reflection. I was not sure if this company was the place for me to be shortly after this discussion. We need to look from four different viewpoints. One viewpoint is from the company's perspective. How would our company see the event or outcome? Another would be from the other person's perspective. How did they see it or understand it? Sometimes we need to get a third parties' thoughts on what happened. Last, we have our own view of what happened. This perspective needs to be humble and seeking to understand. In short, I was in the right place from this reflection. I saw an opportunity to coach up and hopefully impact his thinking. From my reflection, I also found that I should deliver the message up front where we are in the process of improvement and how we are planning to move forward. At the same time, we can learn from even leaders that may not be the best. We can learn from others in their mistakes and we may find out they were not wrong and if we do this reflection, we may see it. The most important part is being humble in your thinking and seek to understand how we can do better as a leader.

Self-reflection is important in how did you handle a meeting. Are your people understanding what you are driving? Are you developing your people? Are you growing in your role? Are you looking for ways to grow yourself? Self-reflection is about having growth-minded awareness. You need to ask these questions about yourself from time to time because we do get caught up in life and forget what we are not doing – growing ourselves and our people. Reflection can help us humble ourselves as well as improve our outcomes in whatever we do in life.

One thing to reflect on as you move up the ladder is, are you connected to the floor or where the work is happening? It is easy to lose this connection because your focus gets wider as you move up. More responsibility, politics in business, and job pressure can change your daily work life. This is one reason you will see the best leaders going to the Gemba (where the work is happening) to understand how the changes they are making affect the processes and the people. By going and seeing, it gives them a chance to reflect.

Reflection is connected to our four philosophies of Lean. By reflecting on your actions, you can see if the results are giving the customers what they want. Going to the Gemba and talking to our people shows respect to them. It improves communication and understanding of how they see the change. Acting on what you have learned from the process of reflection is continuous improvement. We should always have this thought process in both our professional and personal lives. Incremental improvement, putting standards in place to make what we have learned to hold, and looking for the next improvement from what we learn the next time, will drive us all to be better at whatever we do.

Conclusion/Reflection

So why does Lean fail? I believe most would say it fails because the CEO or the boss does not buy into Lean thinking. That is a huge contributor to its failure in most companies and leadership, not buying-in and making Lean thinking its true north, so their people understand the company's vision. Then I have worked at companies where the CEO was bought into Lean as the way to go and I saw Lean failing at different levels of the company. So how can that be if the CEO is talking and walking the way of Lean? Is it because Lean is a way of thinking and it is up to the individual to accept this way of thinking? I would say then that not all of us are on the same level. This can be very hard for some people because they need to humble themselves that they have a lot more to learn. That the MBA and the 20 years in the business may not be enough to be a great leader when it comes to understanding Lean.

Why does Lean fail? We don't communicate well, Lean should be very visual, not just word of mouth or hidden in a computer. We need to understand if we are winning or losing in our day-to-day production. We often don't set our standards high enough. People don't clearly understand what the most important metric to drive is. We don't understand breakthrough thinking and how we can do better than we are currently performing. We need to measure our performance at the process hourly and manage it hourly. The end of the day is too late to adjust. We need a way to signal for help, Andon can be that process. Andon is more than lights and sound. All processes can have an Andon, it's a simple way to send a signal. We don't have an alignment for the vision of where we want to go in the future. Without a clear vision and understanding, there is a great chance that there will be a lack of alignment with your people.

Team leaders and group leaders are a major part of our success. We need to make sure they are developed to solve problems not just firefight a problem. They are closest to our customers when it comes to our product. They see the real change we may make as an organization. We want them to be our future leaders, so their development is very important to our success. As for leaders in our business, we want them to have a servant-leader mindset, to have the ability to be a great coach, so they can lift their team up and develop them to the next level. Our leaders need to have a vision of what we need to be and where they want us to go. They need to have the ability to communicate their message so all can understand it and would then get behind their ideas.

We have no standardized work for what we do. Without a documented process, we have a very poor chance of success over time to be able to have repeatable processes. People come and go; training is shared through word of mouth and little changes happen each time. Without standardized work, we don't have a foundation to understand what should be happening. Every process needs to have some form of standardized work, not just manufacturing processes. People need to know the key points and why they do everything in their process. They need to understand how long it should take to do each part of the process. How often they are expected to complete their process.

We fail at Lean because we may not understand the why for doing something. We need to understand the why in everything we do. When this happens, creativity can grow and move us above what is expected of our customers. The things that people can do are amazing if we set them in the right direction and they understand why they need to do something. Training and communication are key in this along with standardized work. Once people understand the why, they can follow the intent and build upon it. We need to audit and coach not just the standardized work of the process on the floor or office. We need to audit and coach the standards of all our processes. How we have our meetings around our metrics. If you are a leader that visits many sites, do you audit? Do you go-and-see?

We need to develop our people to be creative, to think outside the box, to be experts in what they do, and to be the next leaders in our company. To develop our people, we need them to stretch with our support for them to succeed. It is always good to get cross-functional learning in different departments. Look for the back-office improvements. There is plenty of waste in the office area and we want our people to have a workday filled with value. Use value stream maps to find waste. Value stream mapping can be done all through our business. When thinking about outside the box, we need to look at our supply chain and understand if we can support them to get better. We can build a better partnership with them and create a win, win situation with the relationship.

I don't want us to walk away from this book thinking that leadership is the only reason for failure in Lean. With any journey, change needs to start with yourself. If you want to change the world, start with yourself. Show the world what you do in your day-to-day activities. As I was told by my leader at Toyota when I asked "What is Lean?", it is what we do every day. People should see it in the way our desk looks at work. Are things in order and everything has a place? Do we always look to improve the process that we do? Do we focus on our customers' needs in all that we do? Do we use the go and see approach to seek understanding of what is really happening? Do we seek buy-in from others? Do we show respect for others in what we do every day? Do we communicate well and when we don't, do we seek forgiveness? Are we organized, have a winning attitude, and are humble at the same time?

It is very important that we apply the Lean philosophies and A3 thinking to all that we do. If we do so, it sets us up for success. The Lean philosophies will be in everything that is related to Lean if you look for it. When looking at an opportunity, we should be trying to understand what the standard is, and what is currently happening, so we can truly see the gap that needs to be corrected. Sometimes we try to fix everything and that is not what is being asked by our customers. Most of the time, we may believe we already know the solution to the problem and that thinking makes us fixed-minded. We don't drive to the true root cause and only fix the symptom, hence the problem comes back. The execution of our plan can have some big gaps. We may even fail to measure our success for 90 days to ensure it is truly a success. I have seen many times when we apply pressure to a problem, things seem to get better and that is only because we are giving attention to it at the time. Give it a couple of months and people relax and may revert back to the old way. Then we find how we did not really solve the problem and start to build a mindset that we have tried this so many times and nothing works. We need to use PDCA (Plan Do Check Act/Adjust) in our proposals and our strategy. By using A3, we can all be using a common language and telling our story on one sheet of paper. It can be posted for all to see and understand the intent of what we want to do. A3 thinking is the path to success in working our opportunities.

In everything we do, we need to keep our customers in mind. We need to understand who the customer for each situation is. We need to seek what our customer needs and what would be a delighter/extra that would make our customer feel that we want to serve them. We should always show respect for our people. They are our most valuable resource, only people can develop and grow. We should want to fill their day with value-added work. Don't allow waste-filled work to be the biggest reason they are not with their family. We must always go and see. We need to seek to understand what is really happening, the workplace is always changing. Stand in place, observe the whole, be quiet, and ask why for all that is happening. How could these things you are seeing be better? We need to always be looking for ways to improve what we do. This is in our processes, information flow, movement of people and material, and all the way to how we are as leaders.

Last but not least our reason for why we do what we do. It took a while for us to realize it and we can tell when we discovered our reason for why we do this type of work, it was life-changing to us. We do this because of how we can impact a person's life. We have found many times later on, long after we have worked with someone that they tell the change that they had made because of our interaction. We have seen people cry at report-outs of team leader development. All because of what we had shared with them on how to be a better leader. We get a charge when we return to a site after some time and a team member sees one of us and calls us out with a big smile. There is nothing like a person that has worked for you years ago contacting and explaining how you had a major effect on their life and they want to thank you. It is not about being boastful, it is about explaining that if you apply these philosophies in your daily life, you make changes not just in the KPI or improve company's goal, you can change the world for the better.

INDEX

For Product Safety Concerns and Information please contact our EU
representative GPSR@taylorandfrancis.com
Taylor & Francis Verlag GmbH, Kaufingerstraße 24, 80331 München, Germany

www.ingramcontent.com/pod-product-compliance
Ingram Content Group UK Ltd.
Pitfield, Milton Keynes, MK11 3LW, UK
UKHW052222190926
475702UK00006B/1155
* 9 7 8 1 0 3 2 0 4 8 2 1 5 *